Andrea Wulf was born in Germany as a child. She trai[n]
at the Royal College of Art
Brother Gardeners (longlisted
Johnson Prize and winner of the 2010 American
Horticultural Society Book Award) and *The Founding Gardeners*, and is the co-author of *This Other Eden: Seven Great Gardens and 300 Years of English History* (with Emma Gieben-Gamal). She has written for the *New York Times*, the *Guardian*, the *Wall Street Journal*, the *Los Angeles Times* and many others. She is the Eccles British Library Writer in Residence 2013. She lives in London.

Praise for *Chasing Venus*

'A fine example of scientific storytelling about astronomers of the Enlightenment observing the transit of Venus. Publishers got hot for science writing when *Longitude* by Dava Sobel took off unexpectedly as a long-term bestseller . . . Andrea Wulf's story of how astronomers of the Enlightenment hoped to measure the distance from the Earth to the Sun by observing the transit of Venus internationally on June 6, 1761, and again on June 3, 1769, is another fine example of such scientific storytelling . . . From the original inspiration of Edmond Halley that led to the active co-operation of Captain Cook, Benjamin Franklin and even Catherine the Great, the enterprise is narrated with elegant expertise.'

The Times

'Andrea Wulf's story of the chase is an enthralling, nail-biting thriller and will undoubtedly prove one of the non-fiction books of the year. Even if you fail to see the Transit, don't miss this wonderful book.'

Daily Mail

'Truly excellent . . . rip-roaring tales of the numerous expeditions that set off around the globe to observe the Venusian transit of 1761 . . . communicate[s] the verve and energy – not to mention the perilous nature – of the expeditions.'

New Scientist

'The result is a human story, and it's worth reading as a rallying call to humanity's quest to explore the universe simply for the sake of it.'

Daily Telegraph

'Andrea Wulf's *Chasing Venus* is beautifully paced, alternating between expeditions, with lush descriptions of the often arduous journeys involved.'

Owen Gingerich, *Nature*

'The 18th century stargazers whom Andrea Wulf chronicles in *ChasingVenus* proved themselves a different sort. Their exploits would put Indiana Jones to shame . . . [An] appealing mix of science and travel . . . Wulf writes with enthusiasm . . . and marvelous concision. Better yet, she explains complex scientific phenomena in clear, layperson's terms: Here is a book both astrophysicists and poets can understand.'

Boston Globe

'Entertaining tale . . . tastefully . . . captures the spirit of adventure and the wonder . . . A pleasure to read from beginning to end.'

BBC *Sky at Night* Magazine

'[Wulf's] feeling for personality and her attention to both the scientific records and to the astronomers' journals brings their exploits to life as both scientific exploration and adventurous derring-do . . . enticing tale . . . *ChasingVenus* effectively dramatizes an important moment in the history of science.'

Washington Times

'Andrea Wulf's immaculately researched book describes the endeavours of the early scientific community to observe the transit around the world . . . An absorbing, even exciting yarn.'

The Lady

'Outstanding book! It's the book of the year so far – do not miss it!'

Astronomy Now

'Lively narrative . . . like a non-fiction *National Treasure* with myriads of Nicholas Cages darting around – in a good way. Enlightening Enlightenment fare.'

Kirkus (starred review)

'Andrea Wulf has chronicled the 18th-century transit expeditions in a narrative light on astronomical detail but rich in personalities and adventures . . . she does wonderfully sketch the race for scientific, and patriotic, glory.'

Dallas Morning Star

'[An] excellent book . . . *Chasing Venus* chronicles a rare planetary event that happened at a rare juncture in human history, when the age of empire, the age of science, and the age of curiosity brought the world together for just a few moments – to achieve the measure of the universe.'

Brain Pickings

'Replete with meticulous detail, delightful illustrations and a cast of very familiar names from world history, *Chasing Venus* is an eminently readable account of humanity's effort to chart the heavens. At once an exhilarating adventure, a tale of personal obsession, a tragedy and a detailed history of astronomical endeavour, Wulf's latest work is a fascinating read.'

Press Association

'Thrilling tale . . . a fitting homage . . . an absorbing account . . . Wulf's marvelous eye for detail and talent for simplifying complex science make the book . . . well worth reading.'

Associated Press

Chasing Venus

The Race to Measure the Heavens

ANDREA WULF

 WINDMILL BOOKS

Published by Windmill Books 2013

2 4 6 8 10 9 7 5 3 1

First published in Great Britain in 2012 by William Heinemann

Windmill Books
The Random House Group Limited
20 Vauxhall Bridge Road, London SW1V 2SA

Addresses for companies within The Random House Group Limited can be found at:
www.randomhouse.co.uk/offices.htm

The Random House Group Limited Reg. No. 954009

www.randomhouse.co.uk

A CIP catalogue record for this book
is available from the British Library

ISBN 9780099538325

The Random House Group Limited supports the Forest Stewardship Council® (FSC®),
the leading international forest-certification organisation. Our books carrying the FSC label are
printed on FSC®-certified paper. FSC is the only forest-certification scheme supported by the
leading environmental organisations, including Greenpeace. Our paper procurement policy
can be found at www.randomhouse.co.uk/environment

Typeset in Sabon by Palimpsest Book Production Limited,
Falkirk, Stirlingshire

Printed and bound by CPI Group (UK) Ltd, Croydon, CR0 4YY

To Regan

Coventry City Council	
TIL*	
3 8002 02071 231 3	
Askews & Holts	Mar-2013
523.92	£8.99

'The planet Venus drawn from her seclusion, modestly delineating on the sun, without disguise, her real magnitude, whilst her disc, at other times SO lovely, is here obscured in melancholy gloom'

Jeremiah Horrocks

'We must show that we are better, and that science has done more to humankind than divine or sufficient grace'

Denis Diderot

Contents

Author's Note   xv

Maps   xvi

Dramatis Personae   xxi

Prologue: The Gauntlet   xxiii

PART I   Transit 1761

1
Call to Action   3

2
The French Are First   19

3
Britain Enters the Race   31

4
To Siberia   42

5
Getting Ready For Venus   52

6
Day of Transit, 6 June 1761   68

7
How Far to the Sun?   87

PART II Transit 1769

8
A Second Chance 101

9
Russia Enters the Race 111

10
The Most Daring Voyage of All 122

11
Scandinavia or the Land of the Midnight Sun 133

12
The North American Continent 141

13
Racing to the Four Corners of the Globe 155

14
Day of Transit, 3 June 1769 175

15
After the Transit 189

Epilogue: A New Dawn 200
List of Observers 1761 207
List of Observers 1769 212
Selected Bibliography, Sources and Abbreviations 217
Suggested Further Reading 237
Picture Credits 239
Acknowledgements 247
Notes 251
Index 295

Author's Note

In the interests of clarity and consistency I have retained in the maps and in the text certain place names of the viewing stations as the transit astronomers referred to them in the eighteenth century. Instead of the modern 'Puducherry', for example, I have used 'Pondicherry'; 'Bencoolen' instead of 'Bengkulu'; 'Madras' instead of 'Chennai'; 'Constantinople' instead of 'Istanbul'. In some rare cases where the old names have fallen completely out of use, I have taken the modern name: 'Jakarta' instead of 'Batavia', for example. Please refer to the 'List of Observers' for a full list of the historic and contemporary names.

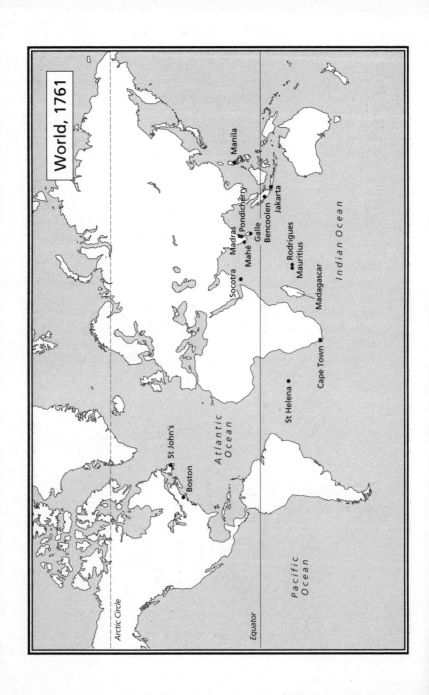

World, 1761

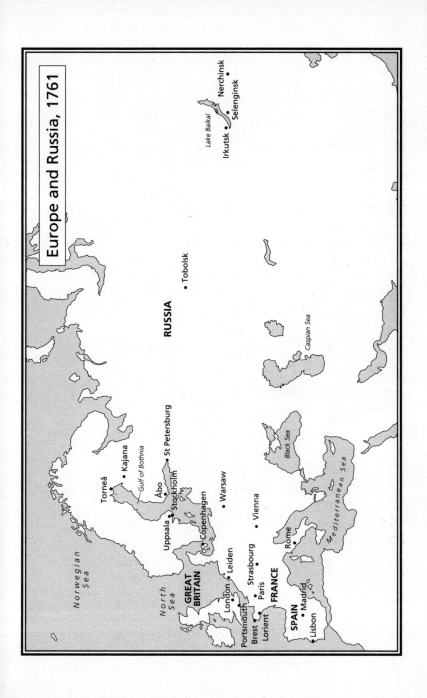

Europe and Russia, 1761

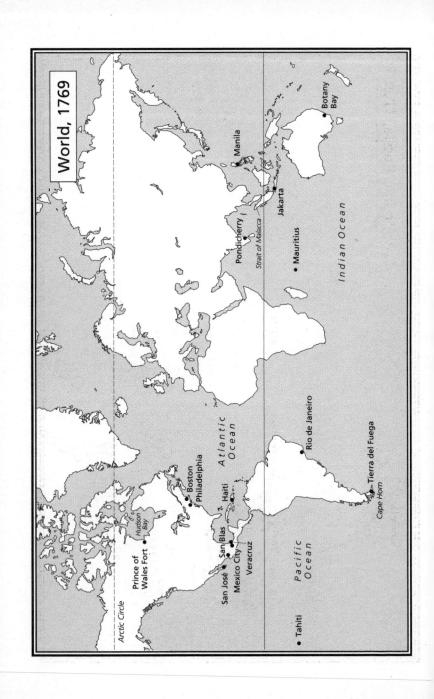

World, 1769

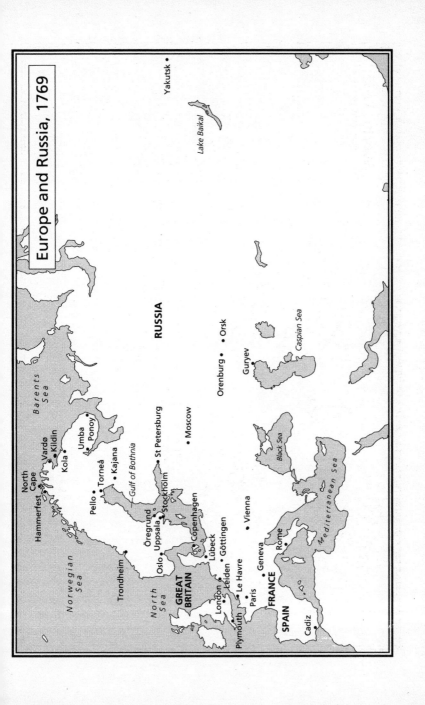

Europe and Russia, 1769

Dramatis Personae

Transit 1761

Britain
Nevil Maskelyne: St Helena
Charles Mason and Jeremiah Dixon: Cape of Good
  Hope

France
Joseph-Nicolas Delisle: Académie des Sciences, Paris
Guillaume Le Gentil: Pondicherry, India
Alexandre-Gui Pingré: Rodrigues
Jean-Baptiste Chappe d'Auteroche: Tobolsk, Siberia
Jérôme Lalande: Académie des Sciences, Paris

Sweden
Pehr Wilhelm Wargentin: Royal Academy of Sciences,
  Stockholm
Anders Planman: Kajana, Finland

Russia
Mikhail Lomonosov: Imperial Academy of Sciences, St
  Petersburg
Franz Aepinus: Imperial Academy of Sciences, St Petersburg

America
John Winthrop: St John's, Newfoundland

Transit 1769

Britain
Nevil Maskelyne: Royal Society, London
William Wales: Prince of Wales Fort, Hudson Bay
James Cook and Charles Green: Tahiti
Jeremiah Dixon: Hammerfest, Norway
William Bayley: North Cape, Norway

France
Guillaume Le Gentil: Pondicherry, India
Jean-Baptiste Chappe d'Auteroche: Baja California, Mexico
Alexandre-Gui Pingré: Haiti
Jérôme Lalande: Académie des Sciences, Paris

Sweden
Pehr Wilhelm Wargentin: Royal Academy of Sciences,
    Stockholm
Anders Planman: Kajana, Finland
Fredrik Mallet: Pello, Lapland

Russia
Catherine the Great: Imperial Academy of Sciences, St
    Petersburg
Georg Moritz Lowitz: Guryev, Russia

America
Benjamin Franklin: Royal Society, London
David Rittenhouse: American Philosophical Society,
    Norriton, Pennsylvania
John Winthrop: Cambridge, Massachusetts

Denmark
Maximilian Hell: Vardø, Norway

Prologue
The Gauntlet

The Ancient Babylonians called her Ishtar, to the Greeks she was Aphrodite and to the Romans Venus – goddess of love, fertility, and beauty. She is the brightest star in the night sky and visible even on a clear day. Some saw her as the harbinger of morning and evening, of new seasons or portentous times. She reigns as the 'Morning Star' or the 'Bringer of Light' for 260 days, and then disappears to rise again as the 'Evening Star' and the 'Bringer of Dawn'.

Venus has inspired people for centuries, but in the 1760s astronomers believed that the planet held the answer to one of the biggest questions in science – she was the key to understanding the size of the solar system.

In 1716 British astronomer Edmond Halley published a ten-page essay which called upon scientists to unite in a project spanning the entire globe – one that would change the world of science forever. On 6 June 1761, Halley predicted, Venus would traverse the face of the sun – for a few hours the bright star would appear as a perfectly black circle. He believed that measuring the exact time and duration of this rare celestial encounter would provide the data that astronomers needed in order to calculate the distance between the earth and the sun.

The only problem was that the so-called transit of Venus is one of the rarest predictable astronomical events. Transits always arrive in pairs – eight years apart – but with an interval of more than a century before they are then seen again.* Only once before, Halley said, in 1639, had an astronomer called Jeremiah Horrocks observed the event. The next pair would occur in 1761 and 1769 – and then again in 1874 and 1882.

Halley was sixty years old when he wrote his essay and knew that he would not live to see the transit (unless he reached the age of 104), but he wanted to ensure that the next generation would be fully prepared. Writing in the journal of the Royal Society, the most important scientific institution in Britain, Halley explained exactly why the event was so important, what these 'young Astronomers' had to do, and where they should view it. By choosing to write in Latin, the international language of science, he hoped to increase the chances of astronomers from across Europe acting upon his idea. The more people he reached, the greater the chance of success. It was essential, Halley explained, that several people at different locations across the globe should measure the rare heavenly rendezvous at the same time. It was not enough to see Venus's march from Europe alone; astronomers would have to travel to remote locations in both the northern and southern hemispheres to be as far apart as possible. And only if they combined these results – the northern viewings being the counterpart to the southern obser- vations – could they achieve what had hitherto been almost unimaginable: a precise mathematical understanding of the dimensions of the solar system, the holy grail of astronomy.

Halley's request would be answered when hundreds of astronomers joined in the transit project. They came together in the spirit of the Enlightenment. The race to observe and measure the transit of Venus was a pivotal moment in a new era – one in which man tried to understand nature through the application of reason.

This was a century in which science was worshipped, and

* Because the orbits of Venus and earth have different inclinations, Venus usually passes above or below the sun (and therefore cannot be seen from the earth). The periods between the pairs of transits alternate between 105 and 122 years. The first transit of Venus observed by an astronomer was on 4 December 1639. The next transits were on 6 June 1761, 3 June 1769, 9 December 1874 and 6 December 1882. There was no transit in the twentieth century but two in the twenty-first – on 8 June 2004 and 6 June 2012. It will be another 105 years until the transit of 11 December 2117.

myth at last conquered by rational thought. Man began to order the world according to these new principles. The Frenchman Denis Diderot, for example, was amassing all available knowledge for his monumental *Encyclopédie*. The Swedish botanist Carl Linnaeus classified plants according to their sexual organs, and in 1751 Samuel Johnson imposed order upon language when he had compiled the first English dictionary. As new inventions such as microscopes and telescopes opened up previously unknown worlds, scientists were able to zoom in on the minutiae of life and gaze into infinity. Robert Hooke had peered through his microscope to produce detailed engravings of magnified seeds, fleas and worms – he was the first to call the basic unit of biological life a 'cell'. In the North American colonies Benjamin Franklin was experimenting with electricity and lightning rods, controlling what until then had been regarded as manifestations of divine fury. Slowly the workings of nature became clearer. Comets were no longer viewed as portents of God's wrath but, as Halley had shown, predictable celestial occurrences. In 1755 the German philosopher Immanuel Kant suggested that the universe was much larger than his contemporaries believed and that it consisted of uncountable and gigantic '*Welteninseln*' – 'cosmic islands', or galaxies.

Humankind believed it was marching along a trajectory of progress. Scientific societies were founded in London, Paris, Stockholm, St Petersburg, and in the North American colonies in Philadelphia, to explore and exchange this new-found knowledge. Observation, enquiry and experimentation were the building blocks of this new understanding of the world. With progress as the leading light of the century, every generation envied the next. Whereas the Renaissance had looked back upon the past as the Golden Age, the Enlightenment looked firmly to the future.

Halley's idea of using the transit of Venus as a tool to measure the heavens was born out of developments in astronomy over the previous century. Until the early seventeenth century man had observed the sky with his naked eye, but technology was slowly catching up with the reach of his ambitions and theories. Astronomy had changed from a science which mapped stars to one which sought to understand the motion of planetary bodies. In the early sixteenth century Nicolaus Copernicus had proposed the revolutionary idea of the solar system with the sun rather than the earth at the centre, and the other planets moving around it – a model that had been expanded and verified by Galileo Galilei and Johannes Kepler in the early seventeenth century. But it was Isaac Newton's groundbreaking *Principia*, in 1687, which had defined the underlying universal laws of motion and gravity that ruled all and everything. As astronomers gazed at the stars, they were no longer in search of God but of the laws governing the universe.

By the time Halley called upon his fellow astronomers to view the transit of Venus, the universe was regarded as running like a divinely created clockwork according to laws which humankind had only to comprehend and compute. The position and movements of planets were no longer seen as ordained arbitrarily by God but as ordered and predictable, and based on natural laws. But man still lacked the knowledge of the actual size of the solar system – an essential piece of the celestial jigsaw puzzle.

Understanding the dimensions of the heavens had 'always been a principle object of astronomical inquiry', the American astronomer and Harvard professor John Winthrop said in the transit decade. Already in the early seventeenth century Kepler had discovered that by knowing how long it took for a planet to orbit the sun, the relative distance between the sun and the planet could be calculated (the longer it took a planet to orbit the sun, the further away it was).* From this he had been able to work out the distance between the earth and the sun relative to the other planets – a unit of measurement that became the basis for calculating comparative distances in the universe.† Astronomers knew, for example, that the distance between the earth and Jupiter was five times that of the distance between the earth and the sun. The only problem was that no one had as yet been able to quantify that distance in more specific terms.

---

* This was Kepler's third law which said that 'the square of the orbital period of a planet is directly proportional to the cube of the semi-major axis of its orbit'. In simpler terms it means that Kepler had provided a mathematical formula which could be used to calculate the relative distances in the solar system by using the radius of a planet and the time it took to orbit the sun.
† The distance between the earth and the sun became the base unit for measuring distances in the universe – it is 1 AU (1 Astronomical Unit). As such the distance between Jupiter and the sun was 5 AU and between Earth and Venus 0.28 AU.

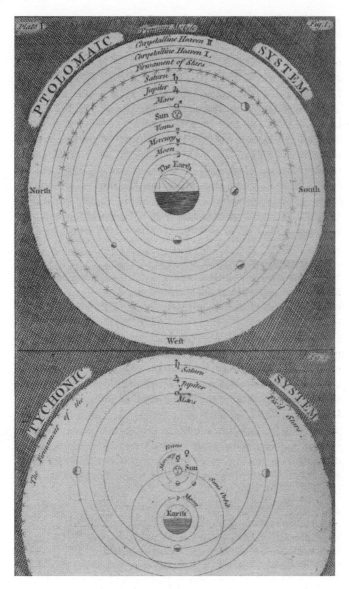

A 1759 depiction of the Ptolemaic and Tychonic planetary systems

Eighteenth-century astronomers had a map of the solar system, but no idea of its true size. Without knowing how far the earth really was from the sun, such a map was all but useless. Venus, so Halley believed, was the key to unlocking this secret. As the brightest star in the sky, Venus became the perfect metaphor for the light of reason that would illuminate this new world and extinguish the last vestiges of the Dark Ages.

Unlike most astronomers whose lives were ruled by the repetitive labour of their nightly observations, Halley had embarked on a far more exciting career – which was probably why he could envisage a global army of swashbuckling astronomers. Not only had he spent one and a half hours in a diving bell submerged almost twenty metres deep in the Thames, he had also undertaken three expeditions to the South Atlantic as the first European to map the southern night sky with a telescope. Halley 'talks, swears, and drinks brandy like a sea captain', a colleague said, but he was also one of the most inspired scientists of his age. He had predicted the return of the eponymous Halley's Comet, produced a map of the southern stars and convinced Isaac Newton to publish his *Principia*.

Knowing that he would not be alive to orchestrate the global cooperation to view Venus's transit – a fact that Halley lamented 'even on his death-bed' whilst holding a glass of wine in his hand – all he could do was to place his trust in future generations and hope that they would remember his instructions half a century hence. 'Indeed I could wish that many observations of this same phenomenon might be taken by different persons at separate places', he wrote. 'I recommend it therefore, again and again, to those curious Astronomers who (when I am dead) will have an opportunity of observing these things'.

Halley was asking his future disciples to embark on a project that was bigger and more visionary than any scientific endeavour previously undertaken. The dangerous voyages to remote outposts would take many months, possibly even years. Astronomers would be risking their lives for a celestial event that would last just six hours and be visible only if weather conditions permitted it. The transit would be so short that even the brief appearance of clouds or rain would make accurate observations difficult or even impossible.

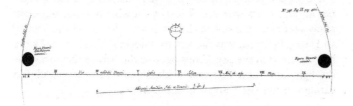

Edmond Halley's drawing of Venus entering and exiting the sun
during the transit

In preparation for it, scientists would need to secure funding for the best telescopes and instruments as well as for travel, accommodation and salaries. They would have to convince their respective monarchs and governments to support their individual efforts and would have to coordinate their own observations with those from other countries. Nations locked in battle would have to work together in the name of science for the first time ever. From many dozens of locations, hundreds of astronomers would have to point their telescopes to the sky at exactly the same moment in order to see Venus's progress across the burning disc of the sun.

And perhaps even more challenging still – though less exhilarating – they must then share their findings. Each observer would have to add his or her observations to the pool of international data. No single result would be of any use without the others. In order to calculate the distance between the sun and the earth, astronomers would have to compare the figures and consolidate the different data into one definitive result. Timings obtained across the world using a disparate range of clocks and telescopes would somehow have to be standardised and made comparable.

The transit of Venus observations were to be the most ambitious scientific project that had ever been planned – an extraordinary undertaking in an era when a letter posted in Philadelphia took two to three months to reach London, and when the journey from London to Newcastle was six days. It took a great leap of the imagination to propose that astronomers should travel thousands of miles into the wildernesses far north and south, laden with instruments weighing more than half a ton.

Their idea of calculating exact distances in space was a bold concept too, considering that clocks were still not accurate enough to measure longitude precisely, and there was as yet no standardised measurement on Earth: an English mile was a different length from a mile in German-speaking countries – which also varied between northern Germany and Austria. A '*mil*' in Sweden was more than ten kilometres, in Norway more than eleven, while a French league could be three kilometres but also as much as four and a half. In France alone there were 2,000 different units of measurement – which varied even between neighbouring villages. In light of this, the idea of merging hundreds of observations taken by astronomers across the world to find one common value seemed outrageously ambitious.

The scientists, who were to leave their observatories in the learned centres of Europe to view Venus from remote outposts of the known world, made for strange adventurers too. At first sight they might not have looked like heroic explorers, but as they chased Venus across the globe they did so with extraordinary intrepidity, bravery and ingenuity. On 6 June 1761 and again on 3 June 1769, several hundred astronomers all over the world pointed their telescopes towards the sky to see Venus travel across the sun. They ignored religious, national and economic differences to unite in what was the first global scientific project. This is their story.

Part 1
Transit 1761

I

Call to Action

By the mid-eighteenth century, at the beginning of the transit decade, the commercial empires of the European countries stretched across the globe. International travel was possible along the established trade routes to distant destinations in the East and West Indies,* Africa and Brazil. Britain controlled much of the eastern seaboard of the North American continent as well as parts of India, some Caribbean islands and Sumatra in Indonesia. France counted among her possessions Canada and Louisiana as well as plantations in India, sugar-producing colonies such as Haiti and St Lucia, and some islands in the Indian Ocean, while the Dutch organised much of their East India trade from Jakarta and ports at Galle in Sri Lanka and the Cape of Good Hope in South Africa.

But voyagers would also face great dangers: since 1756 much of Europe had been embroiled in the Seven Years' War. The political conditions made the transit expeditions perilous enterprises. As scientists from France, Britain, Sweden, Germany, Russia and elsewhere were planning their international co-operation, their armies were fighting bloody battles against each other in the forests of Saxony, on the coast of the Baltic Sea, in the wilderness of the Ohio valley and in India. Rival fleets criss-crossed the oceans from Guadeloupe to Mauritius, engaging in attacks as far away as Pondicherry and Manila but also closer to home in the Mediterranean and the Atlantic.

The war had its origins in the old European conflicts between the Hohenzollerns in Prussia and the Habsburgs in Austria, and

---

* The East Indies as it was known comprised the Indian subcontinent and south-eastern Asia including Indonesia and the Philippines, while the West Indies were the Caribbean islands.

in the ongoing imperial contest between Britain and the House of Bourbon which ruled France and Spain. Britain and Prussia were fighting against France who was allied with Russia, Austria and Sweden. Not only political power was at stake, but also trading and commercial ventures: possession of the North American colonies, of India, the slave trade in West Africa and the valuable sugar-producing islands of the West Indies. As Europeans expanded their world, so did their warfare. It was the first global war – tearing apart Europe and its colonial outposts across the world. It was amid these turbulent times that the astronomers would have to travel on their ambitious quest.

On 30 April 1760, seventy-two-year-old Joseph-Nicolas Delisle, the official astronomer to the French Navy,* walked to a meeting of the Académie des Sciences in Paris. Every Wednesday, academicians who studied in the fields of mathematics and astronomy assembled there to discuss experiments, projects and current research. Delisle only had a short distance to travel. The Académie's rooms were in the Louvre, about a mile across the Seine from his small observatory at the Hôtel de Cluny, the administrative centre of the Royal Navy. The streets were narrow but, as Benjamin Franklin remarked a few years later, 'fit to walk' and kept clean by daily sweeping. They were lined with large houses and busy with people on foot and in coaches. Men and women hawked their wares from stalls – everything from brooms to oysters and from eggs to cheese and fruit. Cobblers, knife grinders and pedlars shouted at the passers-by, offering their services. People 'of all sorts & condition' mingled here, one traveller noted in surprise – from pickpockets to a 'Prince of Blood'. It was, Franklin said, 'a prodigious Mixture of Magnificence and Negligence' – others were harsher and called it the 'ugliest, beastly town in the universe'.

Delisle crossed the river by the Pont Neuf, a sturdy stone bridge famed as the haunt of performers, quacks and tooth-pullers. The bridge was to the city, one Parisian said, 'what the heart was to the body: the centre of movement and circulation'. Turning left, Delisle faced the imposing facade of the Louvre at the next corner.

* Delisle's title was 'Astronome de la Marine'.

France at that time was ruled by Louis XV, a king who had succeeded to the throne in 1715 at the age of five. He adored astronomy, and regularly attended scientific demonstrations in Versailles, even allowing himself to become electrically charged. His great-grandfather Louis XIV had founded the Académie des Sciences in Paris in the previous century to promote science (and its practical uses) and the glory of his reign. Over the past century, academicians had met there to discuss a wide range of scientific subjects, from the study of insects and comets to practical inventions such as hydraulics to power the fountains in Versailles or pumps to clean harbours. The Académie was the most important scientific institution in the country and its members were the best scientists – to be elected as a 'membre de l'Académie' was the greatest scientific honour and the academicians wore their title proudly like a badge of nobility.

The paper that Delisle was about to present would place the academicians at the nexus of the greatest scientific project that had ever been planned. He was going to ask his colleagues to take up the gauntlet thrown down by Edmond Halley forty-four years previously: to set in motion an international collaboration to observe the transit of Venus due to occur one year later, on 6 June 1761.

Halley had propounded the revolutionary idea that Venus's transit could be used as a natural astronomical instrument – almost a celestial yardstick. If several people around the world were simultaneously to watch the entire transit from different places as far apart as possible, he explained, they would each see Venus traversing the sun along a slightly different track – dependent upon the observers' locations in the northern or the southern hemispheres. Venus's path would be shorter – or longer – across the sun according to each viewing station.

With the help of trigonometry, these different tracks (and the differences in the duration of Venus's transit) could then be used to calculate the distance between the sun and the earth. It was an ingenious method because the passage did not have to be 'measured' but only timed – by noting the exact moment of Venus's entry and exit of the disc of the sun. The only equipment the observers would need was a decent telescope with coloured or smoked lenses (to protect against the sun's glare), and a reliable clock.

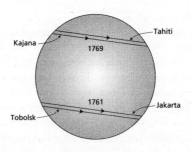

The different tracks of Venus across the sun as viewed from stations in the northern and southern hemispheres during the transits of 1761 and 1769. Locations in the south would experience the longer duration in 1761 and the shorter in 1769.

Since Halley's call to action in 1716, astronomers had tried to find other ways of measuring the solar system. In the early 1750s French astronomers had attempted to calculate the distance between the moon and the earth with observations taken simultaneously from Cape Town and Berlin. By observing the moon from these two locations and with the help of triangulation, they had hoped to measure the heavens prior to the transit of Venus – but the results had not been accurate enough. For years Delisle had believed that he could utilise Halley's method for the more frequently occurring transits of Mercury – he and other astronomers had observed several of those – but he had eventually realised that Mercury was too close to the sun. Only Venus's transit would provide the opportunity to make the calculation.

The task of coordinating the transit observations from many different places on the globe would demand a very particular sort of individual – one so tenacious, persistent and determined that he would be able to unite competitive astronomers and even warring nations. There was no one better suited than Delisle himself. He was an obsessive man, with little time for anything other than the pursuit of science, dedicating his life to the stars. Possessed of an encyclopaedic knowledge and a ferocious work ethic, he was one of the most respected astronomers in Europe. He had worked for twenty-two years in St Petersburg where he had introduced the study of astronomy to Russia, set up an observatory and trained astronomers. He had also turned his journey to Russia into a Grand Tour – not of art and architecture but of scientific men. In London, he had met the ageing Halley in 1724 and discussed the transit of Venus. Now an elderly widower, Delisle was based in Paris and spent most of his time between the Collège de France, where he taught astronomy and lived, and his observatory at the Hôtel de Cluny just opposite.

Not only had Delisle devoted his own life to astronomy, but he also acted as hub for the exchange of information between other members of Europe's scientific community. The volume of his correspondence with foreign astronomers was prodigious, though not everybody agreed with his operating methods. The Swedish ambassador in Paris had been so harried for scientific information without ever receiving anything back that he called Delisle 'greedy'. The French astronomer had the reputation to 'pester all and sundry' for observations but to keep his own a secret. He was 'a devouring gulf which yields back nothing', Jérôme Lalande, one of Delisle's former pupils, complained. Maybe Delisle was sometimes a bit parsimonious with his own results, but he certainly 'devoured' all the information he could about the transit and used his persuasive, if not obstinate, personality to rally the world behind this endeavour.

In the years leading up to the transit, Delisle had studied Halley's astronomical tables, concluding that the British astronomer had been slightly wrong – not in his prediction or call to action but in his choice of the best locations from which to observe the transit. The success of the measurements would depend on making the right choices of viewing stations. As Delisle presented his plan and explained where Venus would appear, his fellow academicians were taken on an imaginary voyage around the world – from Pondicherry in India to Vardø in the Arctic Circle, from Peking to Paris. Halley had predicted that the transit as seen from Hudson Bay on the North American continent would be eighteen minutes shorter than in the East Indies, but, 'I have found', Delisle told his fellow academicians, 'very different results from those of Mr Halley'. According to his own predictions, Delisle claimed that the transit would only be two minutes shorter in Hudson Bay – not enough to aid their calculations – and in any case most of it would occur during the night.

The greatest difference in timings could be achieved if locations in the northern and southern hemispheres would be paired up.

Delisle suggested that Tobolsk in Siberia would be an ideal choice, as would the Cape of Good Hope – the length of duration of the transit as viewed from these positions would differ by more than eleven minutes. To make the selection easier, he also presented a map of the world – his so-called 'mappemonde'. Having originally trained to become a surveyor, Delisle had combined his map-making skills with his astronomical knowledge and produced a map that was shaded in different colours to show where the transit could best be seen. In the blue zone observers would only be able to see Venus entering the sun, in the parts of the world which were coloured yellow only the exit would be visible, but in the red area the entire transit could be seen.

As the scientists examined the map, they could immediately see the best locations, though it also became clear that many of these were far away and would be difficult to reach. The full transit would be visible in China, India and the East Indies as well as near the Arctic Circle, in northern Scandinavia and Russia – with the Siberian location experiencing the shortest transit and the East Indies the longest.

Delisle's presentation to his colleagues at the Académie in Paris was part of a much wider campaign. He had printed his map and explanations of the transit to send them to his international contacts – more than 200 scientists and astronomers in Amsterdam, Basle, Florence, Vienna, Berlin, Constantinople, Stockholm, St Petersburg, and many cities in France.* At the same time French newspapers advertised and described the map, bringing the discussion of the transit into the public domain. Delisle proved to be a worthy disciple of Halley. His mappemonde had been received by every able astronomer in Europe and published in several scientific journals. Delisle could think of nothing else – his apartments at the Collège de France in Paris became the control room of the project and the clearing house for all communications related to it.

Until Delisle asked his fellow astronomers to mount transit expeditions, most of them had lived lives that were an endless round of dull routine: spending cold nights under the open sky

---

* Delisle had distributed almost half of his print run to French astronomers but had also posted around twenty copies to the German-speaking countries, sixteen to Britain, seven to Italy, and several to Sweden, Holland, Russia and Portugal.

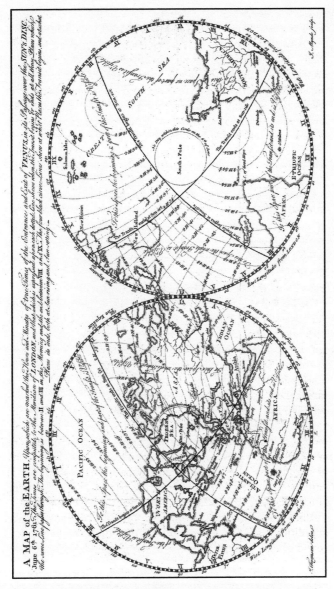

A mappemonde from 1770. Delisle's version would have had regions coloured to represent the visibility of the transit.

or engaging in complex computations.* Though they gazed into the universe day after day and night after night, their own world was rarely extended beyond the confines of their observatories. The only distraction, as one father suggested to his astronomer son, were 'books of voyages' because 'travels would divert and improve'. The job description for the assistant astronomer at the Royal Observatory in Greenwich was depressingly honest: they were looking for men who were 'indefatigable hard working & above all obedient drudges' – not exactly characteristics and requirements normally associated with globetrotting voyagers and heroic explorers.

It was an audacious endeavour, and now, with his petitions sent and little more than a year left before the transit, it was time for Delisle to coordinate the observations and to decide who would be going where. As the commercial reach of the European countries extended across the globe, it made sense to use their existing colonial trade routes in order to travel to the more remote locations. Already Halley had suggested exploiting the imperial possessions of each country, advising the English to travel to Hudson Bay and to India, the French to their plantation in Pondicherry and the Dutch to their trading port of Jakarta – and Delisle agreed.

For the astronomers the transits promised the possibility of scientific revelations and a new understanding of the universe, but they also knew that there were other opportunities presented by the project which they could use to their advantage. If the observers stationed across the world succeeded, the measurements they took would also help improve navigation – essential for any trading empire and naval power. For in tandem with growing empires and Enlightenment ideals, the eighteenth century also became the nursery of capitalism. As new import and export markets mushroomed all over the world, accurate navigation became a branch of science that brought wealth and power. This fact, Delisle was sure, would help convince monarchs and governments to fund at least some of the expeditions.

With the Dutch East Indies as the most distant location

* One of the assistant astronomers in Greenwich summed up what many felt when he wrote 'here forlorn, he spends days, weeks, and months, in the same long wearisome computations, without a friend to shorten the tedious hours, or a soul with whom he can converse'.

and one of the most important viewing stations in the world, Delisle wrote to an acquaintance and fellow astronomer at The Hague in the Netherlands, asking about the possibility of conducting an observation from the Dutch colony. At the same time he continued to canvas support closer to home, begging the French Secretary of State and King Louis XV to pay for a French expedition to Jakarta, pretending that he already had the full cooperation of the Dutch. The gamble didn't quite pay off. Delisle's acquaintance in The Hague had bad news, reporting that the Dutch were only willing to arrange passage for a French observer on a Dutch vessel, but that was all. The Netherlands were unwilling to sponsor any expeditions because 'the usefulness of astronomy to mankind was not sufficiently appreciated in Dutch society', he noted despondently.

Delisle, however, had come up with a solution. As his mappe-monde clearly showed, there were many places where one could observe Venus either entering or exiting the face of the sun. If astronomers used Halley's method of 'duration' (which required the astronomers to see the entire path of Venus across the sun), only a few places across the world would be suitable – many of which, like Jakarta, were far away and difficult to reach. Delisle's new strategy would allow observers to view *either* the entry *or* the exit times of Venus, rather than having to see the entire transit. According to Delisle, an observation of the entry or exit time at one location could be combined with another from a distant location – as long as they were taken in similar latitudes and the exact difference between the places in longi-tude and latitude was known. Astronomers would be able to merge the data after the transit and still calculate the distance between the earth and the sun.

With Delisle as the project's prime mover, it was unsurprising that the French were the first to mount an expedition. On 26 March, five weeks before Delisle had dispatched his mappe-monde across Europe, one of his former pupils acted on his plan and set sail from Brest, a port on the Atlantic coast of France, on his way to India.

Born in 1725 in a little town in Normandy to 'a not very well-to-do gentleman', Guillaume Joseph Hyacinthe Jean-Baptiste Le Gentil de la Galaisière was the first in the race. He had originally pursued an ecclesiastical career in Paris before being distracted by the intellectual stimulation on offer in the metropolis. Once he had heard Delisle lecture on astronomy, Le Gentil turned to science. Instead of praying or getting into 'vain' theological arguments, he now preferred to observe the 'heavens'. He found work at the Royal Observatory in Paris and became a member of the French Académie des Sciences. Like Delisle, he had observed the transit of Mercury in 1753, but had quickly turned his attention to the more useful and rarer transit of Venus, writing about it and then offering to travel to Pondicherry in India where the entire transit would be visible.

At the end of 1759, Le Gentil had received permission to travel to Pondicherry. The combined might of science, politics, and economics – the president of the Académie in Paris, the French Secretary of State and the Controller-General of Finances – were convinced of the importance of the mission and had supported it fully. With promises from the French East India Company, who controlled the trading port at Pondicherry, to provide Le Gentil with passage on one of their vessels, his voyage had been organised within a few weeks. The Compagnie des Indes was, according to Le Gentil, 'always zealous' when it came to 'useful' projects.

Two other French astronomers were also keen to travel: Jean-Baptiste Chappe d'Auteroche and Alexandre-Gui Pingré, who like Le Gentil were also members of the French Académie. Both had volunteered with 'great eagerness' to accept an invitation from the Imperial Academy of Sciences in St Petersburg to travel to Tobolsk in Siberia. It had been decided to send the thirty-eight-year-old Chappe to Russia and the forty-eight-year-old Pingré to another destination, to be agreed upon in due course. Chappe had long been known to Delisle for his precise astronomical calculations and skilful observations and Pingré was one of the most respected astronomers in Paris. Both were regarded as 'worthy' of the honour and the 'perfect' candidates for the appointment – or so at least the members of the Académie thought. They were certainly brilliant astronomers, but also corpulent and middle-aged – not exactly the epitome of daring adventurers. Nevertheless they were ready to face the dangers of the long voyages. France was prepared to chase Venus . . . but Britain was following close behind.

*   *   *

On 5 June 1760, five weeks after Delisle had presented his mappemonde at the Académie in Paris, the fellows of the British Royal Society made their way to Crane Court, a little cul-de-sac off Fleet Street in London, for their weekly meeting. Wealthy fellows arrived in their own carriages, while others walked along the muddy streets or hailed one of the thousands of hackney coaches that choked the narrow lanes. Some called for a sedan chair to be carried quickly through the busy city by porters, who rushed so fast that they often knocked over pedestrians who failed to jump out of their way. They passed the glittering shopfronts on the Strand and Fleet Street. Here, tourists remarked, the shops were 'made entirely of glass' and 'one shop jostles another'. Behind the windows precious wares were displayed, a spectacle of objects that testified to Britain's reach across the globe as well as to her manufacturing prowess. In the evening the flickering light of thousands of candles illuminated shiny silver teapots, political cartoons, new telescopes and heaps of delicate lace. Pyramids of pineapples and grapes competed with diamonds and other precious gems, enticing shoppers to empty their purses.

Every day Londoners were serenaded in the crowded streets by an orchestra of voices and sounds that seemed never to stop – fiddlers playing at the street corners, chimes from the church towers and cries from the street vendors – even during the night they could hear the 'Watchman's hoarse Voice' calling the time and the state of the weather.

When the fellows had climbed the stairs to their meeting room, they excitedly exchanged the latest scientific news and gossip. Their president sat in a large armchair at one end of the long table with a portrait of their royal patron, King George II, behind him, and a marble bust of former president Isaac Newton opposite. As always, it took a moment for all the fellows to settle on the benches and for the chatter to quieten down. Like the Académie in France, the Royal Society was Britain's most important scientific forum. Since its foundation in the 1660s 'for the improvement of naturall knowledge by Experiment', it had become the nexus of British scientific enquiry and Enlightenment thinking. At their weekly Thursday meetings the fellows heard about diving bells and botanical taxonomy, saw exploding dogs, 'electrified' people, and conducted sheep-to-man blood transfusions as well as learning about comets, fossils, and the latest pendulum clocks. Experiments were conducted, results discussed and letters were read that had been received from other scientifically minded people, friends and foreigners alike.

The headquarters of the Royal Society at Crane Court in London

On 5 June, once the attendance had been noted, one of the fellows stood up to read a letter that he had received from Paris, the 'Memoire presented by Mr de Lisle to the Society' and the 'map of the world' that depicted the locations from where to see 'the approaching passage of Venus'. It would set in motion a chain of events which would preoccupy the Royal Society for more than a decade, for when the fellows had finished studying the mappemonde and the transit proposal, they took up Delisle's suggestion enthusiastically.

Only two weeks later, it was decided that the Council of the Royal Society should choose observers and 'proper places' from which to view the transit of Venus. But with only a year left to reach the far-flung destinations – as well as having to organise funds, instruments, and employ astronomers – time was running out. The Council 'unanimously' chose two locations: the remote island of St Helena in the South Atlantic Ocean, the most southerly territory under British control; and a site to be decided upon in the East Indies. The choice lay between Bencoolen (today's Bengkula) on the island of Sumatra, which like St Helena was under the control of the British East India Company, or possibly Jakarta 'if it were not attended with uncertainty', for it was a Dutch possession. In the East Indies the whole transit would be visible while St Helena would only be graced with the exit which, according to Delisle's method, was good enough. The great advantage of St Helena was that it was in the southern hemisphere and therefore the perfect counterpart to viewing stations in the far north.

With the decision made there was a flurry of activity. Some fellows were asked to estimate the expeditions' expenses and to compile lists of the instruments that would be needed. Others were tasked with collecting information on weather conditions in St Helena and the East Indies. Good weather was essential – it would be pointless sending astronomers to the other end of the globe to gaze at a cloudy sky. Most importantly, a delegation was dispatched to enquire of the directors of the British East India Company 'what assistance might be expected from them'.

The Company's involvement was vital. Founded more than 150 years previously as a cartel of merchants who pooled resources to create a monopoly in order to control the supply of goods to their advantage, the Company had gradually expanded. It consisted of a network of colonial outposts that laced the globe, competing with the East India companies of other European countries such as the Dutch or the French. With funds low and timing tight, it made sense to tap into the existing trading network of the empire. If the East India Company proved willing, the Royal Society hoped that astronomers could travel on their vessels, stay in the Company's compounds, and generally make use of the existing infrastructure in these faraway locations.

On 3 July, four weeks after they had read Delisle's letter, the Royal Society's Council reconvened to hear the results of the enquiries: the former governor of Bencoolen had provided the necessary information on the climate there and the meeting with the directors of the East India Company had been a great success, one fellow reported. The directors agreed to do all 'in their Power' to assist the project. There would be no problem, they said, in arriving in St Helena in time. Though it was one of the remotest islands in the world, a lone speck of land in the middle of the South Atlantic, it was an important stopover where vessels replenished their food stores on the East India Company trading route. The voyage would take about three months and commercial sailings were scheduled within this time frame. It would be easy for an observer team to sail on an East Indiaman* and the directors were also happy to provide accommodation in St Helena (though the Royal Society would have to pay for the privilege).

Reaching the East Indies, however, would prove more difficult. There was no company vessel that would reach Bencoolen before 6 June 1761. The directors instead recommended that the Royal Society should contact the Dutch to arrange passage on a ship to their trading port Jakarta, 'which (most likely) will arrive in time'. Meanwhile, the directors had also dispatched letters to their employees in India with instructions on how to observe the transit. After this report another fellow explained that the instruments for the expeditions could not be hired as they had hoped, but would have to be purchased.

Having jotted down all the likely expenses, the Royal Society calculated that they would need a budget of £685 to send an astronomer with an assistant to St Helena, and approximately double that for two observers to go to the East Indies. The cost of the St Helena expedition was almost seven times the yearly salary of the Astronomer Royal and far too much for the Royal Society's small budget – therefore it was decided to write to the Treasury, pleading for funding. Although astronomers across Europe realised the collection of data would have to be a collaborative effort in order to succeed, they also knew that governments and monarchs would be more likely to finance these expeditions if they could be convinced of a national benefit. The Royal Society's petition to the Treasury

---

* An East Indiaman was a vessel that sailed under the charter of the East India Company.

and king appealed to patriotism, and stressed that the honour of the nation should be upheld in this endeavour.

England, the fellows of the Royal Society claimed, had a duty to participate. Not only had the original idea for this project been that of an Englishman, 'Dr Halley, his Majesty's late Astronomer Royal', but the only man who had ever observed a transit of Venus before, had also been an English astronomer – Jeremiah Horrocks in 1639.* More than that, the French and other European nations were about to run away with the prize, the fellows emphasised, for they are 'now sending proper persons to proper places'. The more observations were made, the greater the advantages to science and by extension the participating nations. With the whole world looking to England, the fellows insisted, surely the Treasury would want to answer this 'general Expectation'. For the advancement of astronomy and the glory of the nation, they needed the funds to dispatch their own observers. The strategy worked, and on 14 July, less than two weeks after their petition was sent, the Royal Society received the news that King George II 'had been graciously pleased' to grant the money.

On the same day, without any further ado, twenty-seven-year-old astronomer Nevil Maskelyne was appointed as the principal observer of the expedition to St Helena. The unmarried Maskelyne was a curate in Chipping Barnet, a small town to the north-west of London, but his love for astronomy eclipsed his religious calling. His adoration for the heavens had its roots in his childhood, when he had observed a solar eclipse.† To him astronomical theories were 'sublime' rather than the Bible. Maskelyne had been a fellow of the Royal Society for a few years and had volunteered to sail to St Helena. As part of the voyage was sponsored by the East India Company, it might

* The British astronomer Jeremiah Horrocks had been the first to predict (and see) a transit of Venus. Despite clouds, the twenty-year-old astronomer and his friend, the linen draper and amateur astronomer William Crabtree, managed to observe parts of the transit on 4 December 1639, one in Hoole in Cheshire and the other in Manchester.

† This was the solar eclipse of 1748 which also inspired Jérôme Lalande, a French astronomer who became deeply involved in the transit project. A contemporary said that 'no celestial Phenomonen was ever more useful to Science than this eclipse' because it triggered the love for astronomy in Maskelyne and Lalande.

have helped that Robert Clive was Maskelyne's brother-in-law – for Clive's recent military successes in Bengal had consolidated the Company's ascendancy and eventual dominance in India. For the young amateur astronomer, the voyage was his big chance to step into the wider world of professional astronomy.

Only five weeks after Delisle's letter had been read to the fellows of the Royal Society, the British were ready to stake their claim.

The French Are First

As the fog lifted near the Cape of Good Hope, Le Gentil discovered four ships on the horizon. Some five miles away but closing in quickly, the menacing British warships dwarfed the small frigate on which the French astronomer was travelling. Peering through his telescope, he saw that two of the ships had sixty-four cannons each – the French vessel had twenty-four. The British had been following the ship for the past few days but the weather had always allowed them to escape – until now.

As if sea voyages were not hazardous enough already, the volatile political situation made it all the worse. With the Seven Years' War in full swing, Delisle was sending the astronomers into war zones. Travelling between warring armies in order to reach the transit destinations made for treacherous journeys. With Britain and France fighting against each other, the appearance of the enemy fleet might have meant a premature end to Le Gentil's voyage. Though scientists from the two countries had agreed to work together, the venture was of no significance in the larger political and economic spheres. No matter if the Royal Society in London and the Académie des Sciences in Paris were pursuing the same goal: if a British vessel encountered a French one, they would have to engage in battle. The war had turned seafaring so hazardous that the British East India Company had even told the Royal Society to send two observers to each location but to 'go in *different Ships*' in case one was attacked.

It was not the first time that the thirty-four-year-old Le Gentil had to face the enemy on his journey. Since he had left Brest two months previously, at the end of March 1760, they had been forced to zigzag the ocean in order escape from the British. This time, however, retreat looked unlikely. Le Gentil saw the

British approaching swiftly – despite the strong winds they were under full sail. One vessel was moving to starboard and the other to port, attempting to sandwich the French vessel on the open sea, Le Gentil wrote, and 'to place us between two fires'.

In the face of such danger, Le Gentil displayed an unwavering resolve. He had important astronomical business to attend to, after all, and nothing – neither wars nor waves – would stop him. No matter how tempestuous the oceans or how close the enemy cannons, Le Gentil was ready to risk his life for science and knowledge. That night as they were chased across the rough sea, an unruffled Le Gentil prepared for a lunar eclipse – one of the rare events by which he might determine the ship's exact position. As the earth slowly moved between the sun and the moon, its shadow hiding the moon, Le Gentil pointed his telescope away from the British warships and towards the sky.

Fortunately the weather was on their side, and a thick curtain of fog and rain veiled Le Gentil's frigate from English view, allowing them to disappear into the vastness of the ocean. 'The fog seemed to have been made for us', Le Gentil later wrote, and with the results of his astronomical observations and the lunar eclipse he was even able to assist the captain in navigating the treacherous waters off the Cape of Good Hope.

The strong winds, however, continued to buffet their ship so much that their sails were shredded into useless strips. At least Gentil's seasickness – which had tormented him so badly that he had said death would have been a 'relief' – had abated. He felt well enough to declare that he was feeling 'better at sea than normally on land', taking measurements and observing the stars 'without tiring'. For six more weeks they slowly tacked across the Indian Ocean until they reached Mauritius (then called the Île de France).

Mauritius was a stopover along the French trade route to India and therefore administered by the Compagnie des Indes, and also an important French naval base with a thriving ship-building industry. It was from here that the French launched attacks on British possessions in India and, so Le Gentil had been told, here he would be able to find a ship bound for Pondicherry. Le Gentil disembarked in Mauritius on 11 July – three days before the British Royal Society secured their funding from King George II.

His voyage, Le Gentil now reported rather insouciantly in a letter to the Académie, was 'the nicest and happiest'. They had only lost one man to sickness and 'one passenger who threw himself overboard'. But even Le Gentil, with his talent for sugar-coating the most dreadful situation, despaired when two days later a vessel arrived from India with the devastating news that what was left of the French possessions in India was crumbling under British attacks. Robert Clive's decisive victory at the Battle of Plassey three years previously had already put Bengal under British control. Now Karaikal, a French port just one hundred miles south of Pondicherry, had been taken by the English while Pondicherry itself – the headquarters of the Compagnie des Indes in India – lay under siege. Some 3,000 British, the French captain told a shocked Le Gentil, were 'destined to the siege'. When he had left the Indian coast twenty-five days earlier, the captain reported, the enemy had been busy 'conveying their artillery before Pondicherry'. To make matters worse, much of the French fleet that had been stationed at the naval base in Mauritius and that was to sail as reinforcement to Pondicherry had been destroyed by a hurricane earlier that year – some ships had foundered, others had been smashed against the corals. 'I do not know when I will be able to leave', Le Gentil wrote despairingly to Paris. For the time being he was stuck on Mauritius. It looked as if the first French expedition had already failed.

But Le Gentil was not going to give up that easily and

decided to look for alternative locations from where to view the transit. Tenaciously he tried to concoct a plan, but felt that he was wasting his time on what he called 'chimerical projects'. Running through Delisle's original list of possible locations, Le Gentil first picked Jakarta as a possible alternative to Pondicherry, but eventually gave up the idea. Not one single vessel had arrived on the island while he waited, let alone sailed on to the East Indies. The only option, so Le Gentil decided, was to sail on a small local ship to Rodrigues, an island not too far from Mauritius and known mainly for its turtles. It was not an ideal solution. Le Gentil's calculations predicted that the sun would be very low there during the transit. This would make observations difficult because the horizon was 'always foggy and laden with thick clouds'. Rodrigues's climate in June was not promising because the sky, he was told, was overcast during the monsoon. But there was not much choice, he declared, because 'here I am without hope'.

Besides worrying about his transit observation, Le Gentil's daily life in Mauritius over the next few months was uncomfortable. With Pondicherry besieged by the British, no supplies were arriving from India and the dishonest officials of the Compagnie des Indes in Mauritius sold the goods left in their stores at ridiculously inflated prices. 'Life is horribly expensive', Le Gentil wrote to Paris, moaning in particular about the cost of wine. To make matters worse, Le Gentil was also struck down with debilitating attacks of dysentery. The humid air lay like a thick blanket over the island and he felt weak. His illness, he was certain, resulted from frustration. His 'mortification and concern' about the transit observations had made him sick.

Ironically, as Le Gentil was planning to sail to Rodrigues, the members of the Académie in Paris had decided to dispatch Alexandre-Gui Pingré to Rodrigues as well. With his seemingly unwavering talent for attracting problems, Le Gentil had managed to pick out of the vastness of the ocean the one small speck of land that the academicians had chosen for another French observer. By sheer coincidence two observers who should have been as far apart as possible were about to move towards each other.

*　　*　　*

It had taken the members of the Académie the summer and autumn of 1760 to decide where to send Pingré. During those weeks two French astronomers – with the assistance of Pingré himself – had prepared a report for the Secretary of State and King Louis XV, explaining the importance of the expeditions. Le Gentil's early departure was proof of the Académie's 'zeal', the report stressed, but the French could do more. The transit was one of 'those precious moments', another astronomer appealed in yet another report, and if they didn't use it, they would never be able to make up for the lost opportunity. The previous century had 'envied' them for this moment and the 'future' would blame those who ignored it.

At first, the Académie had hoped to send Pingré to one of the Portuguese or Dutch ports along the south-western coast of Africa such as Luanda in Portuguese Angola or a port in Dutch Guinea. Several locations were considered and evaluated from all angles including Pingré's travel arrangements, weather predictions and the existing infrastructure. The climate in all places, the report concluded, was 'dangerous for foreigners'. They would definitely have to dispatch two scientists, the report's authors added, because if Pingré died, he would 'need to be supplanted'. Bravely Pingré declared that he was 'not alarmed by these dangers' and that they should not consider the 'risks' to his personal welfare.

The forty-eight-year-old Pingré, who suffered from gout, was an unlikely candidate for such a dangerous expedition. His heavy frame and chubby face hinted at his jovial nature and sensuous joy in the good things of life. He was a polymath and ordained priest who had studied and taught theology as well as writing about linguistics, music, poetry and, of course, astronomy. His friendly and lively eyes, though, belied a wilful character. In the past he had incensed his church so much with his unorthodox opinions that they had placed him in an obscure elementary school in the provinces. Bored with life there, Pingré had at the age of thirty-eight turned to astronomy, bombarding the Académie in Paris with scientific letters and essays. As he wrote about comets, eclipses, navigation and the transit of Venus, he had slowly built up his reputation. Finally, critical acclaim for Pingré's astronomical work had even restored his position within the church and he was allowed to return to the Abbey Sainte-Geneviève in Paris, a celebrated seat of learning. Like Le Gentil and Delisle, Pingré had observed the transit of Mercury in 1753 and repeatedly offered his services for the Venus expeditions. With his expertise, the Académie was certain, Pingré's work would 'no doubt surpass' their expectations. It was decided to write to Holland and Portugal to find out which locations were most frequented by their merchant boats and therefore most easily reachable for Pingré.

Unsurprisingly the Portuguese and Dutch, who had not shown much interest in the transit in the first place, were not too keen to provide the French with an opportunity to map their colonial possessions. Their replies politely talked of 'many obstacles'. Quickly, the Académie came up with a new strategy: Pingré would have to observe the transit from a part of the French empire, where he could expect support from the local administration. After some debate, the Académie chose Rodrigues which was part of the commercial web of the French East India Company. The skies were reportedly clear there in June (quite the opposite information to what Le Gentil had heard) and it was in French hands, along the trade route to India.

On 16 November, as Le Gentil lay suffering from dysentery in Mauritius, Pingré joined his friends in Paris for a farewell dinner. The wine was flowing abundantly, the food was excellent and the company cheerful. Only Pingré sat quietly. For once the Frenchman who even in the most adverse situations never lost his appetite, couldn't bring himself to eat. The last few weeks had passed at a frantic pace but suddenly, as he looked at his friends and colleagues from the Académie in Paris, it dawned on him what he was about to do. His companions' chatter faded into the background and he thought of his own uncertain future. Tomorrow he would leave behind the world he knew, travelling in the name of science across the globe. He didn't regret volunteering but he was apprehensive. The appointment, Pingré admitted, had first 'extremely flattered' him, but now his friends' warnings began to trouble him. They were 'the first to be frightened about his fate', Pingré said, and therefore tried to convince him that his life was in danger. Suddenly he saw the voyage through different eyes: instead of fame and honour, death and disease might loom. With the whole of Europe at war, he was risking 'my liberty, my health, and even my life'.

Worried but still determined, the next day Pingré took a coach to Lorient – the headquarters of the Compagnie des Indes – on the coast of Brittany, to board an East Indiaman. On his arrival his fear quickly turned to anger when the local agents of the Compagnie complained that the astronomer had brought a rather excessive amount of luggage. Built as a man-of-war with sixty-four cannons, the *Comte d'Argenson* had been converted into a cargo vessel for the Compagnie. Thirty-eight cannons were removed to make space for commercial goods and passengers and, so Pingré believed, for his astronomical equipment. Outraged, Pingré argued that 700 to 800 pounds of luggage was nothing unusual for an astronomer – the telescopes, quadrant and large pendulum clock were essential. Despite Pingré's protestations, the dispute rumbled on for weeks. The local clerks seemed determined that his adventure was not going to interfere with their rules and regulations. In the end, the Académie in Paris had to intervene and, after weeks of waiting, Pingré finally left on 9 January 1761, with only four months and twenty-eight days until the transit.

With his precious equipment safely stored, Pingré now turned his attention to the second most important issue: food. He questioned the captain about their diet on board, filling his diary with detailed lists of the galley's supplies: cheese, bacon, salted meat, pâté, wine and so on – but discovered to his disappointment that there would be only one meal a day. That first night the sea was rough. Most of the passengers were up all night and 'paid their tribute to the sea'. With his strong stomach, Pingré avoided seasickness but slept badly because of the pain from gout in his right foot. He lay in the dark because they were not allowed to light their lamps or candles during the night for fear of attracting the enemy. His small cabin was separated from the gun deck by temporary partitions – the cannons behind the thin wall were tangible reminders that the seas were a highly contested space, but Pingré was prepared. He carried a passport which the Royal Society in London had procured from the British Admiralty for him, a general order to all commanders of vessels sailing under the

British flag '*not to molest his person* or Effects upon any Account'. He was travelling in the name of science and British captains, so the order read, were to allow the Frenchman 'to proceed without delay or Interruption'.

On his first morning at sea, Pingré heard the sailors calling to each other as they climbed the masts and adjusted the ropes, the wind ruffling the white cloth of the sails. The prow of the heavily laden *Comte d'Argenson* sliced through the grey depths, leaving behind a trail of white foam like an ephemeral tail in the waves. As the captain bellowed orders, a cacophony of loud voices, running feet and clinking metal echoed through the ship. A fleet of five British warships had been spotted only two or three miles away, preparing to attack. As the captain commanded his men to the cannons, they tore down the new partition walls. Luggage, timber, ropes and cannonballs tumbled into chaos. Where their cabins had been, passengers now saw heavy artillery pushed into position. In a last attempt to avoid battle the captain veered the ship and for hours they zigzagged across the sea with the British behind them. Whenever they thought they had put some distance between themselves and the enemy, yet more British ships appeared. Between noon and 7 p.m., Pingré counted eight new vessels. Then night fell and the wind suddenly changed and they finally managed to draw away into the darkness. 'Providence', he wrote that night in his journal, had decided that they escaped 'without firing a shot'.

After this lucky retreat the journey became relatively uneventful. Occasionally they saw enemy vessels in the distance, but they always managed to avoid a fight. In the early weeks when his enthusiasm for the voyage had not yet been dampened by the idleness, Pingré enjoyed the music and dancing of the sailors which had turned the whole ship into 'a grand ballroom'. With each day merging into the next without anything happening, monotonous ennui took hold of Pingré. He ate, observed the night sky and fished. Usually jovial and sociable, the astronomer declared himself bored by the other passengers who were mostly employees of the Compagnie des Indes. It would be preferable, he said, to be 'alone than in the company of people one doesn't like'. There were not enough books, too much noise and no space to walk. So dull was life on board that another passenger said he would rather be imprisoned in the Bastille.

Only occasionally were these seemingly elastic days interrupted by something new. One morning, for example, Pingré discovered that according to their charts they had sailed across land at the Cape Verde islands and joked with the crew that the *Comte d'Argenson* was 'such an excellent vessel that it could split land and rocks with the same ease as the waves of the ocean'. They saw flying fish, sometimes the sea seemed to be 'on fire' with phosphorescence, and once a sailor had to be rescued from the ocean after a fall from the mizzenmast. The most memorable day was their equator crossing. The old sailors prepared the 'equator baptism' for days, dressing up as the so-called 'le père *la Ligne*' – the 'father of the Equator' – and practised pranks that they could play on those who had never crossed the line before.

And though the 'ceremony' and the jokes were silly, there was something majestic about this moment, especially for an astronomer. Once they had sailed across the equatorial line, they entered the southern hemisphere with a sky that displayed a glittering dome of stars that Pingré had never seen before. The astronomer who had previously said that 'liquor gives us the necessary strength to measure the distance between the sun and the moon' began to regard his observations more seriously, now taking his measurements 'not with the bottle, but with the octant'.

Everything went according to plan until 8 April 1761, just after they rounded the Cape of Good Hope. In the morning they saw a ship in the distance and feared it was the enemy – but it turned out to be the *Le Lys*, a French supply vessel that had been attacked by the British. The *Le Lys* was packed to the brim with provisions from the Cape for Mauritius to fill the stores of the Compagnie des Indes but damaged so badly that the captain ordered Pingré's vessel to accompany and protect them.

As they slowly sailed towards Mauritius – with the *Comte d'Argenson* reducing sail cover to adapt her speed to that of the *Le Lys* – Pingré grew increasingly exasperated. They would never make it to Rodrigues in time if they had to stop in Mauritius first, he said. His voyage would be 'completely useless'.* He argued, begged and pleaded, and even threatened the captains with legal proceedings. One evening, he wrote a formal letter of complaint, reminding both men that he was travelling in the name of the French king, the Académie des Sciences and the Compagnie des Indes, with explicit orders to sail to Rodrigues. It was 'the holiest of his duties', Pingré wrote, the 'whole of Europe' was watching because his observations were important not only for France but for science. When this pompous outburst failed to have its desired effect, Pingré tried to convince both men with rational thought and logic as he calculated their exact position again and again, explaining to the captains that they were actually on course for Rodrigues. At first the captain of the Le Lys had tried to appease Pingré with fresh fruit and meat from the Cape, but eventually he became so annoyed with the astronomer's incessant grumbling that he threatened to 'throw him over board'.

Pingré's calculations of their route had been correct. On 3 May 1761 he saw Rodrigues on the horizon. Though they were so close, for him the island had become unreachable. One of the officers on the Comte d'Argenson made a last attempt to convince the captain to stop briefly and let Pingré disembark. But there was no point in pleading any more, the captain had made up his mind. Pingré passed Rodrigues on course for Mauritius. After a four-month sea journey, he missed his destination by only a few miles.

Had Le Gentil's plan of leaving Mauritius in order to observe the transit in Rodrigues suceeded, the two French astronomers might have crossed at sea. As it was, surprisingly, Le Gentil had managed to find passage to India. At the end of February a ship had arrived in Mauritius from France with orders to send reinforcements to India to relieve the besieged Pondicherry. After some deliberations, the island's governor and commander-in-chief decided to dispatch the Le Sylphide, one of the few ships left in Mauritius's decimated fleet. Le Gentil immediately saw his chance. The timing was tight but the sailors reassured

---

* Pingré had also heard that Le Gentil was stranded on Mauritius (probably from the Le Lys' captain) which made his own presence on the island doubly obsolete. With one observer already in Mauritius, there was no point in his going there too.

him that they could make it in two months – just in time for the transit. Le Gentil didn't need any further convincing. Pondicherry was, according to Delisle and Halley, one of the most important locations on the globe and Le Gentil's calculations for viewing the transit at Rodrigues and Mauritius were not promising. If the town was still in British hands, Le Gentil was sure he would be able to find another viewing station on the south-eastern coast of India.

On 11 March 1761, Le Gentil left Mauritius and, after a short stopover at the nearby island of Réunion (then called Île Bourbon), they set course for Pondicherry. At first, an exuberant Le Gentil noted, everything seemed to go smoothly. Each day the ship covered between thirty and forty-five miles, proceeding swiftly until they hit the north-east monsoon, north of Madagascar. Instead of being able to sail in a diagonal line across the Indian Ocean towards India, the winds pushed them towards Africa. They made little headway. Because April and May mark the transitional period of the monsoon, when the winds begin to turn and become the south-west monsoon during the summer months, Le Gentil woke each day, praying that their direction had changed. But his prayers were in vain. Instead of encountering strong westerly winds that would have carried them to Pondicherry, the vessel was enveloped in a lull, as if someone had suddenly pulled the brakes.

At the end of April, Le Gentil could see to his dismay the coastline of the island of Socotra just to the east of the Horn of Africa. It was only one month to the transit but he was still some 2,500 miles away from Pondicherry.

The sails hung motionless from the yards and the ship seemed to stand still on the mirrored sea. By day the heat charged the air with a trembling shimmer, and in the evenings as the sun set Le Gentil stared across the ocean, imagining that 'golden sequins' had been sprinkled on the endless expanse of the water. The spectacle roused his senses. He had never seen anything like it and thought the sun's rays resembled 'golden columns' that reached from the horizon to their ship. No cloud ever tainted the perfect blue cupola above them, but no wind billowed their sails either. As spectacularly beautiful as this was to look at, their vessel was not making progress. Having left Mauritius seven weeks previously, they were still closer to Africa, than they were to India.

Then in mid-May, with less than a month to the transit, they finally hit the south-west monsoon. As the ship sliced its way through the ocean, Le Gentil allowed himself to hope again. By the end of the month, he could see a line of lights glimmering in the distance. They had been fast but not fast enough – the lights were not those of Pondicherry, on the south-eastern coast of India, but of Mahé on the south-western coast, which was also a French trading port. They now had less than two weeks to make it to Pondicherry.

The next morning, when they sailed closer to the coast, Le Gentil saw the English flag flapping in the wind. Two small boats stopped his vessel, presenting letters from the governor of Mahé. He wrote that the port had been taken by the British. The news grew worse still as Le Gentil read on: Pondicherry also had fallen to the siege. There was now no hope of viewing the transit from there.

'To my great vexation', Le Gentil wrote in his journal on 24 May, it was decided to return to Mauritius, no matter how hard he begged. With no time to lose and the oceans infested with British ships, the captain shouted orders to change course. And as if the heavens and the sea had also turned against them, they then faced a storm so mighty that their vessel was lifted high on to the crest of the waves. Le Gentil could not fathom what was happening. He had been the first astronomer to leave Europe and now, more than one year later, so close to his goal, his chance of observing the wondrous planetary encounter had been snatched away from him by the British Army.

With Le Gentil's hopes dashed by the redrawing of the imperial map and Pingré having sailed past Rodrigues, two of the most important transit expeditions seemed to have ended before Venus had even come close to the sun.

3

Britain Enters the Race

As Le Gentil and Pingré criss-crossed the oceans, the British were finalising their plans for the expeditions to Bencoolen in Sumatra and St Helena in the South Atlantic. Curate Nevil Maskelyne had made sure that he would be equipped with the best instruments but also that he received a suitable liquor allowance – the bill for wine and spirits accounted for almost one-quarter of the entire budget for the expedition. The East India Company had offered transport and accommodation in St Helena but with no commercial sailings scheduled that would reach Bencoolen in time, the Royal Society had to come up with another idea. They needed a vessel, so where better to go than to the Royal Navy? The fellows approached the British Admiralty to seek their assistance.

Exact navigation was crucial to any trading empire and seafaring nation – it brought wealth and power. Arguing their case carefully, the Royal Society reminded the Admiralty that the promotion of science was 'intimately connected with the Art of Navigation'. Not only would the transit of Venus help astronomers, but since Delisle's method required the knowledge of the exact geographical position of the viewing stations, the observers would also be creating a web of accurately measured locations.

At the time of the transit, very few places across the world had been determined precisely. Even the exact difference between the longitude of the royal observatories of Greenwich and Paris was yet to be established. In Russia the first atlas of the empire based on scientific methods had only been published fifteen years previously – until then not a single Russian city had been accurately mapped. Sea charts were also notoriously unreliable – something that Pingré had discovered for himself at the Cape Verde islands. Although ships regularly called at ports in this region along the European trade route to Africa and the East Indies, the charts were so imprecise that Pingré's ship had – according to their maps – sailed across two of the islands. The charts urgently needed to be updated, Pingré had noted, which was exactly what the Royal Society was recommending with their proposal.

At sea the lack of a precise knowledge of longitude could have disastrous consequences. Even the most experienced captains lost their bearings and whole squadrons had disappeared due to miscalculations. Sailors died, ports were missed (and with them fresh water, fruit, vegetables and other important provisions). Since latitude was easily calculated (by measuring the height of the sun above the horizon at noon),* vessels tended to sail along a straight line of latitude, crowding the vast oceans along narrow paths – like busy highways in a desert – thereby making themselves easy targets for pirates and enemy vessels.

Latitudes are the belts slung horizontally around the globe – and in parallel to each other – with the equator marking the zero-degree, while longitudes are the lines that run from the North to the South Pole. The globe is divided into 360° longitude which represents the twenty-four hours that it takes the earth to rotate – so 180° marks twelve hours, 90° marks six hours, 15° is one hour, and 1° is four minutes. Calculating longitude was easy, theoretically. By knowing the exact time in the home port and the local time in the current position, the difference between these times could be translated into a geographical position – or the difference in longitude. If it was one hour later than at home, the distance from the home port was 15° to the east, if it was two hours later one had travelled 30° towards the east – and if it was two hours earlier one had travelled 30° towards the west, and so on.

---

* Latitude is calculated at noon by finding the angular distance between the sun and the observer's position – this is the angle of the observer's position and the sun against the horizon. Simplified, this means that the observer can find his latitude by subtracting this angle from 90°. If the angle is, for example, 70° then the latitude is 20° and if the angle is 35° then the latitude is 55° and so on. At the equator with the sun directly above the angle is 90° and the latitude is 0°.

The local time was easily set at noon, when the sun reached its highest point, but knowing what time it was in the home port was altogether more complicated. There were two ways to work it out: to travel with a clock set at the time of the home port or to observe a celestial event that was predicted to occur at a specific time at home (and then to compare it with the local time). The movement and position of the moon against a background of fixed stars could also be used, as could eclipses of Jupiter's satellites or of the moon. A lunar eclipse that was predicted for ten o'clock in the evening in Greenwich, for example, provided a sailor who observed it at two o'clock in the morning with the knowledge that he was 60° east from Greenwich.

There were two problems: firstly most astronomical events were too rare to be of any practical use; and secondly there was no clock that would keep time reliably on board a rolling ship. In 1760, the most precise astronomical timekeepers were pendulum clocks – great on land but completely useless on a pitching boat where they would slow down, run fast or stop altogether. Fluctuating temperatures thinned or thickened lubricating oils and expanded or contracted the metal parts of the clock. For trading empires the ability to calculate longitude was so important that, in 1714, the British government had already set aside the enormous sum of £20,000 (200 times the yearly salary of the Astronomer Royal) for the Longitude Prize – for the discovery of such a method. In Paris and in Greenwich observatories had been established with the express purpose of finding longitude by studying the skies – but it remained a task so elusive that Jonathan Swift had equated the quest with the discovery of perpetual motion and a 'universal medicine' in *Gulliver's Travels*.

The Admiralty was therefore quick to respond to the Royal Society's request. Only a week after the initial contact they 'ordered a Ship to be fitted for the said purpose'. Though the transit observations had not been organised with the specific goal of discovering a way to determine longitude, the expeditions were nonetheless recognised as potentially useful to that purpose. Ambitious Nevil Maskelyne even planned to use his voyage to St Helena to test a new longitude method based on lunar observations with which he hoped to win the coveted Longitude Prize.

The Royal Observatory at Greenwich

With colonial commerce becoming increasingly significant to the British economy, the support of the Admiralty and the East India Company for the transit expeditions came as no surprise. Since the accession of George II, in 1727, exports to the West Indies had more than doubled, and to the East Indies they had grown ninefold. The members of the French Académie had successfully used the same argument when they appealed for funds from their government. With about two-thirds of French foreign trade involving colonial products, navigation was an essential pillar of the country's economy. Mastery of the colonies depended on the mastery of longitude – and that in turn depended on the mastery of astronomy (at least for as long as there were no clocks that would work precisely on board a ship).

Nevil Maskelyne and his assistant left Britain for St Helena at the end of January 1761, just as Pingré's ship sailed past Madeira and as Le Gentil was agonising about where to go to from Mauritius. Maskelyne's ship was accompanied by several heavily armed vessels which were sailing to the West Indies. At the Canary Islands the convoy turned west while Maskelyne's ship continued south. Over the next few weeks he perfected and tested his method of determining longitude at sea. Night after night, with the glittering Longitude Prize at the back of his mind, he peered through his telescope, measuring the moon's path across a tapestry of fixed stars. Maskelyne noted the local times of the moon's wanderings and then used so-called 'lunar tables', lists that predicted when the moon would pass which star at a particular place in Europe – which then acted as his zero-degree longitude reference point. Comparing these times with local times allowed Maskelyne to calculate the difference in longitude – if the moon passed a particular star in Greenwich, for example, at 2 a.m. and Maskelyne saw it at 1 a.m., he knew that he had travelled 15° westward.

Over the previous century astronomers in Greenwich, Paris, Nuremberg and elsewhere had gazed into the night sky to compile a map of the stars and the moon's slow march against them. The Royal Observatory in Greenwich had been founded with the explicit goal of creating such a map 'of the Motions of the Heavens' in order to calculate longitude. The first Astronomer Royal John Flamsteed had taken around 30,000 observations and only recently the German astronomer Tobias Mayer had finished the first lunar tables that gave the moon's position every twelve hours. Mayer's lunar tables were so revolutionary that he had submitted them to the British Board of Longitude in order to claim the Longitude Prize for himself.

As they sailed south, the night sky became Maskelyne's heavenly clock. If the night was clear, he was on deck to measure the distance between the moon and the fixed stars with a quadrant. On page after page, row after row, he noted his observations and calculations, tightly squeezed on to the paper. Then he referred back to Mayer's lunar tables which the Board of Longitude had given him in order to test. 'My principal attention on board', Maskelyne later told the Royal Society, was 'to be satisfied . . . of the practicability of that method'. The only disadvantage of the lunar method was that it involved such complicated calculations that sailors couldn't glance quickly into the sky to work out their longitude. Each of the calculations was so complex that the whole process took around four hours, which was no problem for Maskelyne who adored lists and order, but was unsuitable for practical naviga-tion. Throughout the journey the astronomer patiently explained everything he was doing to the officers on board, knowing that the Longitude Committee might call for witnesses to confirm the viability of the method. He was pleased to report that a person who took accurate observations in this way and had the ability and 'leisure' for the long calculations, could 'ascertain his Longitude as near as will be in general required'.*

It was not all hard work on board ship. Maskelyne travelled with more than 100 gallons of wine and rum as well as five gallons of spirits and over seventy bottles of claret. It was, he noted, 'a very agreeable voyage'. The ocean crossing combined the things he loved: good wine, astronomical observations and plenty of time to fill his notebook with long lists of measurements, longitudinal calculations and reports on the weather. In fact, so preoccupied was he with his work that he didn't even stop when St Helena finally appeared on the horizon.

View of Jamestown, St Helena

The morning of 5 April 1761 dawned clear with a gentle breeze. In the distance rose the dark and barren cliffs of one of the most isolated places in the world† – 1,200 miles from Africa and 1,800 miles from South America, in the midst of the South Atlantic – a black jagged rock which still displayed the violence of its volcanic origins. Another traveller described

---

* Upon arrival in St Helena the ship's officers who had used their old method were up to 10° wrong, while Maskelyne was only 1.5° out.
† St Helena was so remote that the British thought it was the perfect location for Napoleon's exile. He arrived on the island in 1815 and died there almost six years later.

how the cliffs 'seem almost to overhang', frighteningly close as ships approached. It was a dangerous manoeuvre. Measuring ten miles by five, the island was the home of only a few hundred people and had been in the possession of the British East India Company for nearly a century.

Early the next morning, with exactly two months until the transit, Maskelyne wrote in his journal: 'came to an Anchor in the Bay before James's Fort St Helena'. After a journey of eleven weeks and two days, they had arrived at Jamestown, the only town and harbour on the island. It was time to find a suitable place to set up an observatory.

At the same time as Maskelyne had sailed towards St Helena, another British team had set out for Bencoolen, the East India Company trading port on Sumatra which Halley and Delisle considered as one of the most essential locations for viewing the transit. For this journey the Royal Society had selected thirty-one-year-old Charles Mason and twenty-six-year-old Jeremiah Dixon. Mason worked as an assistant at the Royal Observatory in Greenwich, a job that had taught him the intricacies of astronomy and the use of all the latest instruments. Dixon was an amateur astronomer and surveyor from the north of England, who had probably been recommended by one of the fellows who was his neighbour there.

Like the zealous Maskelyne, both men saw the expedition as an opportunity to improve their professional standing, but there was also the lure of fame and adventure, and an escape from the monotony of an astronomer's life. Occupied seven days a week with long and laborious calculations, Mason also had to get up three to four times a night to observe the skies, no matter how bad the weather. His job in Greenwich was so dreary and lonely that 'nothing can exceed the tediousness and ennui', as one assistant complained. Equally frustrated, Dixon seemed to have drowned the dullness of his life in alcohol and had just recently been expelled from his Quaker meeting house for 'drinking to excess'. Another temptation to join the expedition might well have been the stipend. Both men were to receive £200 each for their services (and £30 for provisions and liquor), quite an improvement on Mason's yearly salary of £26 at the Royal Observatory. There was little else keeping them in Britain.

Dixon was unmarried and Mason's wife Rebekah had died the previous year.

Once again the East India Company was asked to provide information. One of their captains was invited to a Royal Society meeting where he was quizzed about the local weather, workmen and materials available at Bencoolen. The Court of Directors of the East India Company promised to do 'every thing in their power for facilitating the making of observations' by sending instructions to the governor of Sumatra and by providing passage, workmen and food, 'all at the Companys Expence'.

No other observers would travel as far. To reach Bencoolen, Mason and Dixon had to sail from Portsmouth on the south coast of England to Spain, then along the west coast of Africa and around the Cape of Good Hope, before traversing the wide expanse of the Indian Ocean. Although they had boarded HMS *Seahorse* in late November 1760 – on the same day that Pingré left Paris – they had been delayed for weeks by contrary winds. They finally sailed on 6 January 1761, but only four days later, at exactly the same time as Pingré's ship was chased by the fleet of British vessels off the coast of France, the Seven Years' War brought Mason and Dixon's voyage to a violent halt. Like Pingré and Le Gentil, they were about to discover just how dangerous it was to be caught in-between warring nations. At eight o'clock in the morning, as the rising sun was erasing the night, the outline of a solitary frigate appeared behind their vessel. When the sky brightened, the captain discovered that it was a thirty-four-gun French vessel, 'crouding down upon him'. It would be an unequal battle – HMS *Seahorse* had only twenty-four guns and was difficult to manoeuvre because she was heavily laden for the long voyage to the East Indies.

There was no chance of escape. They would have to fight. Within two hours the French were close enough for Mason and Dixon to see their faces – they were within 'pistol-shot'. Soon enough, battle began. The smell of gunpowder filled the sea air, but in the ensuing chaos of shots and screams, Mason and Dixon couldn't tell who had the upper hand. Their mast came crashing down, hit by a French cannon, and another was badly damaged. Splintered wood, torn sails and jumbled ropes covered every surface. Suddenly the first Frenchmen were standing on the deck of the *Seahorse*, attacking her crew. The planks which had been scrubbed clean in Portsmouth a few days earlier were now stained with blood. As the fighting raged back and forth between the French and the British, Mason and Dixon were certain that their lives were over.

Though the French had at least double the number of men, the British were not giving up easily. Pressing forward, one by one, they regained control of their ship, driving the French back to their own vessel. At noon, after a battle that had lasted little more than an hour, the French retreated, their commander dead along with half of his crew. Yet there was little cause for celebration on the English ship. Eleven men lay dead and forty-two were wounded, 'many of whom', Mason noted, 'I believe mortal'. As the ship's surgeon dealt with the casualties, and the captain inspected the damage, Mason and Dixon opened their trunks which had been tossed around during the fighting. They had brought with them two reflecting telescopes, a so-called micrometer which would allow them to measure the diameter of Venus and the sun, a quadrant to calculate the altitude and positions of planets as well as a large astronomical pendulum clock. Taking out each instrument, they found everything in good order, with only the stands broken. The *Seahorse* hadn't been as lucky, and with her sails and rigging torn and every mast damaged, the captain decided to sail to Plymouth to have the 'shattered' ship refitted. Only days after they had begun their adventure, Mason and Dixon had to admit that it would be 'absolutely impossible' to reach Bencoolen.

Traumatised by the battle, weakened by seasickness, and worried that they might have to take the blame for a failed observation, the two men panicked, bombarding the Royal Society with letters in which they insisted that they would not be able to fulfil their contract. As these letters were read out at an emergency meeting of the Royal Society, the fellows began to suspect their two explorers of cowardice. The two once-eager astronomers, the horrified fellows learned, now categorically refused to go to Bencoolen – seemingly unconcerned that the expedition had been financed by the Crown, supported by the Admiralty and organised by the Royal Society itself.

According to their own calculations, Mason and Dixon now indicated, the best place for them to view the transit would be from Scanderoon in Asia Minor, in the most north-eastern region of the Mediterranean.* Though they said that they would 'obey' the commands of the Royal Society, their promises rang hollow. In the same sentence they threatened that they would 'not proceed from this to any other place' – at no other location, Mason and Dixon explained, would they be able to 'perform what the world in general reasonably expect from us'. And in the vain hope of sympathy, Mason added a postscript which stated that he had been suffering from 'continual Sickness' while at sea. On the same day he posted a similar letter to his former boss at the observatory, the Astronomer Royal, who was also a fellow of the Royal Society, writing that he saw 'no reason why I should go upon impossibilities'. His behaviour might seem 'strange', he admitted, but that didn't stop him from declaring that they wouldn't go anywhere else 'let the consequence be what it will'.

Mason and Dixon had clearly misjudged their position. The fellows were outraged – to them this was mutiny and they unanimously resolved to draft a letter that ordered the two disobedient astronomers 'to go on Board the Sea Horse, and enter upon the Voyage'. They were 'surprised' by the pair's behaviour, reminding Mason and Dixon that they were bound by their contract. Their unwillingness to depart for Bencoolen, so the fellows warned, would not only harm the nation and the Royal Society but would also prove 'fatally to themselves'. Other countries were looking to Britain, they told their recalcitrant astronomers, and defiance would only cause a 'scandal' and 'end in their utter Ruin'. To make matters even clearer, the fellows ended their letter with the threat that any further refusal would be met with 'inflexible Resentment' and land them both in court.

Meanwhile the captain of the *Seahorse* was similarly baffled, writing to the Admiralty and the Royal Society that Mason and Dixon 'absolutely refused to proceed the Voyage'. Confused as to whose decision it was – the astronomers' or the Royal

* Scanderoon was the ancient city of Alexandretta and today's Iskenderun in Turkey, where Asia Minor meets the western coast of Syria.

Society's – the Admiralty cancelled the voyage. The secretary of the Royal Society, Charles Morton, dashed back and forth to the Admiralty, trying to undo the damage. He was furious. So far, with only Nevil Maskelyne on his way to St Helena, Britain lagged behind in the global endeavour.

After all, the British believed that the French were planning three major expeditions: Le Gentil's voyage to Pondicherry, Pingré's to Rodrigues and Chappe d'Auteroche's to Russia. Moreover, only a few days previously, a letter had reached the Royal Society which detailed the 'preparations now making in Sweden'. Apparently, the secretary of the Royal Academy of Sciences, astronomer Pehr Wilhelm Wargentin, had been busy over the past months, organising viewing stations in cities such as Stockholm, Uppsala and Lund. Following Delisle's suggestions that locations in the far north would provide essential data, Wargentin was also planning to send astronomers to Torneå in Lapland and Kajana in eastern Finland (then also called Cajaneburg or Cajaneborg). He had already established nine locations, and was working on more.

Worried about Britain's contribution, Morton went to the Admiralty to clarify the situation. Mason and Dixon's 'peremptory refusal', he explained, was no reason to end the enterprise because they were going to Bencoolen, whether they wanted to or not – otherwise they would be punished as mutineers 'with the utmost Severity of Law'. Nothing, it seemed, was to stand in the way of the Royal Society's plans. Once they had received notification of this, Mason and Dixon realised that further protest was pointless and reluctantly complied with the order to sail for Bencoolen.

'We are sorry', they told the Council of the Royal Society, that suggestions of travelling elsewhere had been interpreted as mutiny. 'We shall to our best Endeavours make good the trust they have pleas'd to confide in us', they wrote, and left Britain once again.

4

To Siberia

As astronomers in France, Britain and Sweden were setting off, willingly or otherwise, the Imperial Academy of Sciences in St Petersburg was struggling to find observers. Though Delisle had spent twenty-two years in St Petersburg, introducing astronomical studies to Russia and founding the observatory, there were still not enough trained astronomers in the country. The problem was not new. In 1725, when Peter the Great had established the Academy, there had been so few Russian scientists that he had had to staff it entirely with foreigners. Even now, almost forty years later, less than half the members of the Academy were Russian – the majority were German.* Foreign scientists had been enticed to make the long journey to Russia with double salaries in a bid to bring knowledge to the empire. In general, though, Russian science was rather looked down upon – much of what was published in Russia, as one German scientist wrote, was 'brazenly derided' elsewhere in Europe.

When the Russians had heard that the French were sending Le Gentil to Pondicherry, the secretary of the Imperial Academy in St Petersburg (a German) had 'commended' his colleagues to contribute to the global endeavour with a transit observation of their own in Siberia. But the response had been unenthusiastic – and so they had written to their former colleague Delisle in Paris to ask if the French

---

* The numbers of German academicians were dwindling as Russia battled Prussia in the Seven Years' War, but afterwards during the reign of the German-born Catherine the Great, more German scientists arrived. In fact, there were so many that the minutes of the Academy meetings were written in German during that time.

would be willing to assist. The Académie was keen to help, and recruited astronomer Jean-Baptiste Chappe d'Auteroche for the task. At the end of November 1760, as Pingré set out on his voyage to Rodrigues, Chappe left Paris for Siberia.

As the son of a baron, thirty-eight-year-old Chappe didn't need to make a career in professional astronomy. Since childhood he had been passionate about mathematics and the stars, impressing many with his grasp of the most complicated calculations. As an adult he had spent most nights gazing at the sky as well as accepting the occasional commission to survey land – if the clients were suitably aristocratic. He had produced only a few scientific papers – and those at a leisurely pace* – and in the previous year he found employment at the Académie in Paris.

The sedentary life of an astronomer was not for Chappe – the Venus adventure was much more to his taste. The dangerous journey across war-torn Europe was not for financial reward but for the sake of science, and glory. 'He liked fame', his colleagues noted. The Siberian observations, Chappe was certain, would add a new chapter to man's knowledge of the universe because Delisle favoured Tobolsk as one of the most important locations. The entire transit would be visible there and it would occur for the shortest duration – making it the perfect counterpart to the longer transit to be seen in Bencoolen in the East Indies.

It would be an exhausting 4,000-mile journey from Paris to Tobolsk, and its beginnings did not bode well. When Chappe left Paris it rained incessantly, turning the roads into channels of deep mud. By the time he drove into Strasbourg eight days later, his carriage was already beyond repair. He had to purchase a new one as well as replace his thermometers and barometers which had been damaged in accidents – though his telescopes emerged intact. Frustrated by the bad roads, Chappe decided to make the next part of the journey by boat along the Danube, but progress was again hampered by the weather. Thick fog made it unsafe to travel except during the

---

* Chappe had translated parts of Halley's astronomical tables to French, observed the transit of Mercury in 1753 and published a few papers in the Academy's journal.

day, and then often 'only for a few hours'. All the while the celestial clock was ticking.

Despite these setbacks, Chappe remained cheerful and entertained himself with drawing maps, noting local customs and calculating the distances between towns. He also found time for other valiant occupations. Near Regensburg in Bavaria, he rescued a man who was about to throw himself into the Danube in a suicide bid 'on account of a quarrel with his mistress'. The next day, after a stroll through a small town, Chappe found on his return a melancholic fifteen-year-old girl on the deck of his ship. Quizzing her, Chappe discovered that she had run away from her uncle who wanted to force her 'to take the veil'. Gallantly, Chappe delivered her back to her parents.

When he reached Vienna, Chappe swapped his boat for a carriage and decided to travel through the day and night. He bought lights so that the coachman could see at least the most dangerous potholes and rocks on the road, but accidents became so frequent that Chappe was 'in continual apprehensions for my instruments'. As the roads grew worse, wheels smashed, and on 10 January – the fateful day that Mason and Dixon's vessel was attacked by the French – Chappe's carriage crashed into a ditch somewhere on the road between Brno and Nový Jičín, in today's Czech Republic.

It was bitterly cold and snow dusted the empty landscape. The hills through which the road led disappeared in the darkness. Chappe stood in the ditch pulling and pushing his carriage. With him was the coachman, the new secretary of the French embassy in St Petersburg (who had joined him in Vienna) and several servants – all desperately trying to rescue the battered vehicle while icy water seeped through their shoes and clothes. However hard they shoved and dragged, their combined strength could not shift the carriage. They finally agreed that they would have to unload their luggage, and placed the leather bags stuffed with clothes and small trunks filled with papers on the road. When they lifted the heavy wooden trunks which contained the precious astronomical and meteorological instruments, Chappe became agitated. Hearing the clinking sound of glass, he knew that once again something had broken. That night Chappe, who was usually gregarious and optimistic, allowed himself for the first time to contemplate defeat. 'I began to fear', he wrote in his diary, that 'we should not reach Tobolsky in due time'.

Without the heavy equipment in the carriage, they eventually freed it and continued their slow journey towards Russia. While his colleague Le Gentil was suffering the humid heat in Mauritius, Chappe endured a cold he 'had not before experienced'. Even inside the carriage the temperatures were so low that one day he fumbled out his thermometer with numb fingers and scrawled in his journal 'eleven degrees below 0'. He had to wade waist-high through sluggishly floating ice when the carriage crashed through the frozen surface that had transformed the rapid rivers into temporary roads – probably no surprise given that the instruments alone weighed more than half a ton. The hilly roads on land posed other problems as they were covered 'from top to bottom' with an icy glaze. Even when they put all ten horses in front of one carriage, they found they couldn't move it. Much of the way through the mountains they had to walk, slipping and falling, and were soon covered in bruises. Sometimes strong winds blasted clouds of snow high up into the air, whipping the flakes into frozen pellets. The coachman, who was most exposed to the frosty assault, 'could not stand it' and ran away.

In Riga they switched to sledges only to discover half a mile outside the city that the snow had disappeared. They were stuck once again. Their interpreter – trying to keep warm – was of no use for he was 'in liquor' as Chappe found out. Then, just before they reached St Petersburg, their carriage slipped into such deep drifts that only the heads of the horses poked out. They were 'buried' for hours.

Chappe was frustrated by these 'continual delays', yet he managed to find some distraction from the misery by examining the women he met along the way with the taxonomic precision of a scientist. He compared, examined and categorised. In one village he measured their petticoats and in another declared the ladies 'strictly virtuous'. No matter how cold or exhausted he was, he remained a connoisseur of the female sex and remarked appreciatively on their sparkling eyes, the 'slenderness of their waists', and 'well-shaped servant maids'.

The Imperial Academy of Sciences and observatory tower in St Petersburg

On 13 February, almost three months after he had left Paris, Chappe finally arrived in St Petersburg. He now had to prepare for the final leg of his journey. It was another 1,800 miles to Tobolsk, but first he was introduced to the members of the Imperial Academy of Sciences and presented with an essay on the transit of Venus, written by one of the academicians. The essay itself was at the centre of a vicious row which had been rumbling on for weeks between two of the most distinguished astronomers in the Academy. Russian scientist Mikhail Lomonosov was at war with his German colleague Franz Aepinus who, according to Lomonosov, had been unfairly promoted over his Russian colleagues. While Lomonosov had been finding it hard to gain support for his own scientific projects, Aepinus, a foreigner and newcomer, had been fast-tracked up the gilded ladder of imperial favour so quickly that he had recently been made the personal tutor of the future Empress Catherine the Great.*

Lomonosov was a brilliant scientist but was feared for his explosive temperament. The ardently patriotic son of a fish trader from Archangel had been the first Russian admitted to the Imperial Academy in St Petersburg. He was a polymath – writing about language, poetry, history, art, chemistry and

---

* Aepinus's career continued to flourish and in early 1761 he was appointed Director of Studies at the Imperial Corps of Noble Cadets – a position that brought not only prestige but also a much larger salary.

astronomy. Lomonosov hated the many foreign scientists at the Academy because, he said, they were not doing 'any service to the Russian fatherland'. His colleagues, the poet Aleksandr Pushkin later wrote, 'dared not utter a word in his presence'. Not a man who bowed to polite etiquette, the battles he had waged against his fellow academicians were so violent that Lomonosov had once even been put under house arrest for eight months after a drunken brawl ended in a stabbing. He would use any tactic to challenge his enemies, resorting to smear campaigns, slanderous newspaper articles and illicit backroom dealing. Lomonosov insulted, bullied, and enjoyed disturbing the Academy's meetings by playing indecent pranks on his foreign colleagues. In his eyes they were stupid 'lapdogs' at best, and Machiavellian rogues at worst – and were all, he was convinced, conspiring against him.

When Aepinus had first published his essay about the transit of Venus, Lomonosov thought it simplified the astronomical principles to such an extent that it was plainly wrong. The argument had grown into a full-blown dispute, with Lomonosov composing his own essay about the transit in response as well as writing letters of complaint to his fellow academicians. Aepinus's drawings were incorrect, he claimed, the terminology unscientific, and the treatise would be of no help to any observer. At the same time Lomonosov tried to undermine his adversary's credibility: the observatory, he claimed, was rarely used, and the laboratory which was Aepinus's responsibility was a mess with tools coated in mildew and rust. Only recently, Lomonosov told the president of the Academy, had he gone to the observatory during an astronomically important night and had found the entrance covered in snow – Aepinus had clearly not bothered to visit.

Aepinus retorted that nothing was wrong with his essay and that Lomonosov was spreading 'false rumours throughout the city'. In vain he hoped that this would end the battle. But Lomonosov was only just beginning. Aepinus's scheme, he said, was 'flawed'. It wasn't even clear for whom he had written his essay: the 'rude and uncultured mass' would never understand it, while the intelligence of noblemen and the academicians would be 'insulted' by the plain text. Aepinus used the academicians – who 'hate me', Lomonosov told his colleagues – to incite even more disputes. Before Aepinus continued with his 'wanton quarrels', the angry Russian growled, he should bear in mind Lomonosov's 'services to his country' and not treat him like an amateur. It was ridiculous. The two astronomers who should be working together in preparation for the transit were at loggerheads.

Chappe was not about to get involved. He had problems of his own, because he discovered during the meeting that he had astronomical competition. At the end of November, just as he had struggled through those torrents of never-ending rain, the Russians had decided to mount their own expeditions. One observer was to travel to Irkutsk near Lake Baikal in Siberia, and another to Nerschinsk near the Chinese border. It seemed that the letter which the Académie in Paris had written, informing the Russians that they were sending Chappe to Siberia, had never arrived. With no news from France, the president of the Academy had announced that Russia needed to participate, echoing the French and British sentiments that the transit observations were important for the science of navigation as well as for the 'honour' of the empire. They had procured the necessary instruments and had trained two young observers for the task.

The Russians had set off a month before Chappe arrived in St Petersburg, fighting their way towards the east through the frozen desert. With their own observers on the way, the Russian Academy now regarded Chappe's expedition as pointless. The academicians argued that he should view the transit somewhere more convenient and closer to St Petersburg, but the Frenchman thought otherwise. He had not endured the hardships of his long journey to be ouflanked by the Russian scientists. Highly competitive and keen to provide the most important data of all, Chappe began to rally support for his Tobolsk expedition, arguing that there was no other part of the globe where the transit could be observed 'to so much advantage'. He went to the French ambassador in St Petersburg who 'easily comprehended' Chappe's reasoning, and to the High Chancellor of Russia who was luckily a 'lover and protector of the Sciences'. After four weeks in St Petersburg Chappe had achieved what he had set out to do and was granted permission to continue his journey.

One of Chappe's enclosed sledges in which he travelled from
St Petersburg to Siberia

On 10 March 1761, with less than three months until the
transit, Chappe left St Petersburg late in the evening. He travelled
with an entourage of four enclosed sledges, each drawn by five
horses that carried all he needed for the observations in Tobolsk.
At night the sledges were illuminated by torches, the lights flickering between the trees, a long train of strange shapes and heavily
breathing animals moving through the darkness. For the journey
through Siberia, Chappe was obliged to take even the most basic
provisions: food, of course, but also beds and wine. One sledge
was loaded with his instruments, while another carried his servant
as well as his watchmaker – Chappe had decided that he needed
someone 'to mend my clocks in case of accident'. The third
sledge was occupied by a sergeant who had joined the expedition
as a guide and interpreter and in the fourth was Chappe himself,
wrapped in furs and staring out on to the blanket of snow that
reflected the moonlight with a ghostly shine.

Nothing would stop him from reaching Tobolsk, he promised
himself, but it would be a race against the thaw. Chappe prayed
for the freezing temperatures to hold. He would never make
it in time on the rough roads – the only option was to travel
on the frozen rivers. As much as he hated the cold, the Arctic
embrace was essential to his success. As the sledges flew through
the winter landscape, Chappe experienced for the first time the

delights of this mode of transport. 'We went with the greatest velocity', he exulted. Tobolsk, he felt, was waiting for him.

For part of the journey they travelled on the Volga river, across a surface as 'smooth as glass', the sledges gliding with 'inconceivable swiftness'. Chappe was so invigorated by the speed that he climbed out of his cabin to feel it properly, standing 'upright' on the sledge's roof with the icy wind blasting around him. His joy did not last because accidents continued to delay them. The sledges constantly overturned, a horse almost drowned, they hit overhanging trees, and fell again and again into deep snowdrifts, Chappe said, 'liable to be swallowed in the snow'. But he still found time to continue his survey of Russian women, noting those who were 'lively', 'taller', 'very pretty', had 'better complexions' or 'a very disagreeable figure'.

The interior of a Russian cottage according to Chappe – complete with a half-naked woman

At the beginning of April, four weeks after he had left St Petersburg, the ice began to melt, cracks opened up and water began to seep through the frozen surface of the rivers. Determined not to be 'overtaken by a thaw', Chappe perilously raced on. On 9 April, the ice had become so thin that his coachmen refused to cross the last river. They were only fifty miles away from Tobolsk – a twelve-hour sledge ride – but if they failed to cross now, their destination would remain as unreachable as Venus herself. Without the assistance of the governor of the city, Chappe would neither be able to build an observatory nor would he have any protection in the Siberian wilderness. He needed the infrastructure that Tobolsk would provide.

After all that he had endured, Chappe refused to believe that the expedition had come to a halt so close to its goal. 'A cold sweat came all over me', he despaired, 'attended with an universal dejection'. He cajoled, threatened, bantered and bullied, trying to talk his companions into taking the risk. When he finally convinced them, with the help of copious amounts of brandy, it was night and so dark that they could only see the faint glimmer of the stars reflected in the treacherous ice. Although water had started to seep out over its thawing surface, Chappe hurried his drunken men along. Scared but determined, he stood on top of his sledge, driving his small expedition through the water across the thin ice.

5

Getting Ready For Venus

With the transit less than two months away, astronomers across the world were busy preparing for Venus's rendezvous with the sun. But many of the expedition astronomers had yet to reach their destinations, and even those who had arrived still had a great deal of work to do – from erecting their observatories to establishing their longitude. Closer to home, scientists in Britain, Sweden, France, Germany and Italy also began to get ready for the auspicious day. All those observing in central Europe would miss the first part of the transit during the night, but would still be able to time Venus's exit at first light. Delisle's call to action was answered a hundredfold. Across Europe observatories were being erected and carpenters were busy making viewing platforms. In Munich, the members of the Bavarian Academy of Sciences used the occasion to build the city's first observatory (albeit only a small one, in a tower on the rampart in the Hofgarten). A monastery in Germany ordered instruments from France and positioned them on its garden walls, and a Polish Jesuit put his telescope on the top gallery of a library in Warsaw. In the Netherlands astronomers constructed viewing towers on their houses and in Austria noblemen turned their castles into temporary observatories. Astronomers in Rome, Vienna, Göttingen, Amsterdam, and in dozens of other towns in Europe checked their instruments, set the clocks and waited for Venus.

The Swedish Royal Academy of Sciences was also busy with its preparations. The Academy had been established in 1739 – during the so-called Age of Freedom, when power shifted from Crown to parliament following the death of King Charles XII in 1718. Sweden, a once-great imperial player, was seeing

An astronomer observing the transit of Venus

its influence in the imperial tug of war decline. The country had been embroiled in some disastrous wars, including the ongoing Seven Years' War, in which they had allied with France against Prussia. The ruling party in parliament, the so-called 'Hats' (named for the headgear worn by officers and gentlemen), sought closer links to France and focused their interest on mercantile policies and the economy. Unsurprisingly – with several Hat politicians as founding members – the Academy had been initiated with the specific goal of encouraging 'useful' sciences. Any subject that had the potential to be profitable for Sweden's economy was encouraged: from astronomy and new agricultural tools to 'silkworm farming'.

The secretary of the Academy, forty-three-year-old Pehr Wilhelm Wargentin, the greatest astronomer in Sweden, was determined not to be outdone by his colleagues elsewhere in Europe. The son of a pastor with a penchant for science, he had observed the sky as a boy, and later studied astronomy at the University of Uppsala. Wargentin was dedicated to astronomy, and had transformed the Academy in Stockholm over the previous two decades from an inward-looking domestic organisation to a thriving scientific forum that contributed to the lively international exchange of ideas. He had long believed that 'a kind of brotherhood' should exist between the different scientific societies across Europe, and had established contacts with many astronomers and scientists in Sweden, as well as with his colleagues abroad. Wargentin's thirst for exchange and cooperation was unmatched by almost any of his contemporaries. The amount of work he did – administrative and astronomical – was so prodigious that he had no time for anything else. He might not have been amusing or sociable, but he was efficient, a punctual correspondent, and much admired by his peers. He was Sweden's most important link to the international world of science.

'Before my time,' he said, 'the Academy had (so to speak) no correspondence with foreign countries', but now the publications of the Academy were translated into German and summarised in French. In exchange for Swedish journals, Wargentin received similar publications from the societies in London, Paris and St Petersburg. Organising the transit observations was part of his endeavour to encourage these closer links. All Swedish astronomers, Wargentin insisted, should contribute, with the Royal Academy of Sciences as the organising body behind their efforts.

Delisle had sent a mappemonde to his friend Wargentin, encouraging him to contribute to the global initiative. Sweden was in an advantageous position because Scandinavia was the most easily accessible region in the world where the entire transit would be visible. Not only that – as it was important for the calculations to be made from locations as far apart as possible, the viewing stations in the far north were the perfect counterpart to Nevil Maskelyne's observations on St Helena in the southern hemisphere. As a result Wargentin had ordered some instruments from the best craftsmen in London, and had organised an expedition to Lapland in the northern part of Finland (then a Swedish possession) where the White Nights of the north would allow the observers to see the whole transit. He was the driving force behind the Swedish contributions.

The Swedes, like the British and French, made use of their colonial possessions to conduct the transit observations. Sweden might have lost its status as a major imperial power in the early eighteenth century, but the vast emptiness of Lapland with its small population of reindeer-herding Sami was for the Swedes as much of a colonial outpost as Jakarta and Sumatra were for the Dutch and British. As such, Lapland was as exotic and unknown to them as a tropical island in the Indian Ocean, and promised – so many Swedish scientists believed – as many riches.*

For the Lapland expedition Wargentin had asked astronomer Anders Planman, who worked as a teacher at the university in Uppsala, to travel to Kajana in eastern Finland (then in Swedish hands). The eager Planman was closely involved with all the preparations, suggesting an innovative design for a twenty-five-foot telescope that could be dismantled into six parts for easier transportation, and designing the packing boxes so that even the 'thinnest' lenses could be safely carried on a farmer's cart. Planman left Uppsala in the middle of the winter, in February 1761.

---

* The botanist Carl Linnaeus, for example, had travelled to Lapland in 1732 and suggested establishing plantations in the mountains above the treeline.

To reach Finland, he had to cross the frozen Gulf of Bothnia by sledge, but the severe winter had laid an unusually thick blanket of snow over Scandinavia. The whipping waves had congealed into a frosted picture, as if someone had snapped a finger to stop the world. In place of a smooth surface the Gulf of Bothnia was a treacherous icescape of 'superb stalactites of a blue green colour'. Though stunningly beautiful, it made for dangerous travelling. Sledges had to follow the hardened lines of the waves, regularly overturning when one side would suddenly be 'raised perpendicularly in the air'. Wrapped up in thick pelts, the passengers were often catapulted out of their sledges like furry cannonballs and the horses then galloped off, scared, as another traveller described, 'at the sight of what they supposed to be a wolf or bear rolling on the ice'.

By the time Planman arrived in Åbo on the Finnish side of the Gulf of Bothnia, he was so ill that he was forced to rest for three weeks until he recovered his strength. To make up for lost time, he then travelled day and night through lonely forests towards Kajana. There was a 'dreary silence', other travellers remarked, the only noise the erratic choir of bursting bark which exploded with a bang when a tree's sap froze and expanded. With the snow more than a metre deep, the horses became stuck so regularly that Planman walked for large parts of his journey. He shared the nights with strangers, horses, pigs and dogs in simple log cabins, but enjoyed every passing day that brought him closer to his destination. He arrived in Kajana on 15 April, two months after his departure from Uppsala, and just days after the British astronomer Maskelyne had disembarked at St Helena.

While Planman fought his way through the snow, there was one group of scientists who were disappointed that they were not given a chance to participate in the global enterprise. The thirteen British colonies in North America would experience the entire transit in the hours of darkness. As Venus began her journey across the sun, colonists from Georgia to Massachusetts would be fast asleep. By the time the sun rose in the morning, she would have disappeared again. But there was one American who was determined not to miss the sight: forty-six-year-old Harvard professor and astronomer John Winthrop. He was so desperate to catch at least a glimpse of the planet, that he was willing to travel to the easternmost tip of the North American continent: Newfoundland. There, with luck, he would be able to see the last minutes of Venus moving across the face of the sun.

He had left his planning late. In early April 1761, Winthrop published a long article in several American newspapers, explaining the importance of the event. As each of the thirteen colonies had their own government, Winthrop appealed to his local assembly in Boston, emphasising the commercial aspect of the project which 'may ultimately be very serviceable to Navigation'. It was a wise approach. Trade powered the American colonies – their fields fed the mother country and in return the colonists imported much of their goods from across the Atlantic, while Newfoundland's fisheries depended on their export markets in Europe and the West Indies. Two weeks later, on 20 April, Winthrop had the permission and financial support of the government of Massachusetts, which was now convinced that a transit expedition would 'do Credit to the Province'. They would provide a vessel to take him to any location in Newfoundland which he 'shall judge proper'. He had less than seven weeks to organise the expedition, reach his destination and set up an observatory.

As the day of the transit drew closer and astronomers became nervous, newspapers picked up on the story too, inciting enthusiasm among amateurs and the general public. In April 1761, the *Edinburgh Magazine* explained in great detail the intricacies of the transit. The article reported on the British expeditions to St Helena and Bencoolen and praised the French efforts to send astronomers to Pondicherry and Rodrigues. Only the Dutch, the author continued, had shown no interest – they seemed more concerned with 'carrying on an illicit trade with the French, than willing to send observers to Batavia'. In the American colonies the *Boston Evening Post* reported that King George II had granted money for expeditions to Bencoolen and St Helena and in Stockholm Wargentin wrote a long article about the transit for a Swedish newspaper with instructions on how to view it. Underlining the importance of the rare event, he also provided information about instruments and how to protect the eyes from the dangerous glare of the sun using smoked glasses. French newspapers ran articles about

Venus, and Delisle distributed a pamphlet called *Vénus passant sur le Soleil* about the forthcoming observations.

Books were published for amateur observers, including a whimsical 'dialogue' between a brother and a sister that explained the transit to young gentlemen and ladies. A Scottish astronomer and instrument-maker called James Ferguson translated Edmond Halley's treatise 'for those who do not understand Latin', and instructed the reader on how to use a quadrant to measure distances between planets.* He also added a translation of a French essay on the transit which had been published by the Académie des Sciences in Paris, a map which was based on Delisle's mappemonde, and engravings of Venus's path across the sun. His lectures on astronomy for the general public, Ferguson said, had become so fashionable that he could easily have filled the seats twice. Benjamin Martin, an English instrument-maker, was also taking advantage of the new-found excitement for astronomy, offering entertaining talks on the transit at his shop in London's Fleet Street. Day after day, in the weeks before the transit, he placed announcements in local newspapers, advertising his services. Everybody who was willing to pay two shillings and sixpence could learn about Venus with the help of Martin's fascinating array of orreries, globes, diagrams and maps. The subject had become so popular that he too wrote a book, called *Venus in the Sun*.

Meanwhile the professional astronomers tried to keep each other updated as much as possible. Pamphlets were exchanged, news conveyed and predictions shared. The Royal Society informed Delisle that Mason and Dixon's ship had been attacked by the French only four days after their departure but that they had left for Bencoolen (or so it was thought). The chaplain of the British community in St Petersburg updated the fellows of the Royal Society with an account of the Russian expedition, as well as sending them Franz Aepinus's offending essay on the transit. A Dutch astronomer promised Delisle that he would do all he could to find an employee of the Dutch East India Company to view the transit in Jakarta since the

* Ferguson had presented his book *A Plain Method of Determining the Parallax of Venus by her Transit over the Sun* to the Royal Society on 19 February 1761.

During the transit decade astronomy became a popular pastime

Dutch government had refused any official cooperation. When Chappe had been introduced to the members of the Imperial Academy in St Petersburg, he had been able to tell them first-hand about all the French expeditions. A member of the Bavarian Academy of Sciences in Munich informed a colleague that he had heard from Pingré about his voyage to Rodrigues, and the German secretary of the Russian Academy had written to a scientific friend at the Royal Prussian Academy of Sciences in Berlin with details of Chappe's journey, which were then reported to a colleague in Leipzig.

News of the far-flung expeditions slowly trickled into the capitals of Europe. The French scientists received a letter from Le Gentil written in July 1760 from Mauritius. For all they knew, he was now on his way to the East Indies because he had notified them that 'I have planned to go to Batavia' (today's Jakarta). They also read a letter from the Russian Academy which announced that Chappe had left St Petersburg for Tobolsk – but they had as yet no word of him reaching his destination. The Royal Society did not know that Maskelyne and his assistant had disembarked at St Helena, and had no idea what had happened to Mason and Dixon.

They could only hope that everything was going according to plan. And to some extent it was. Chappe had won his race against the thaw when he safely crossed the last river. On reaching the other side the astronomer had collapsed 'with an universal tremor'. But he had made it and, on 10 April 1761, four days after Maskelyne set foot on St Helena, Chappe arrived in Tobolsk – a week later the ice broke up and melted, bringing the most severe spring floods the region had ever experienced.

By mid-April 1761 only Maskelyne, Chappe and Planman had reached their destinations, but for the other expeditions a successful observation seemed an unattainable goal. Mason and Dixon were sailing along the western coast of Africa, thousands of miles from Bencoolen, Le Gentil was still in the middle of the Indian Ocean and Pingré was on his way to Mauritius instead of Rodrigues. Even those who had arrived, still needed to build their observatories, set up their instruments and determine their longitude by observing the stars – time was running out.

Maskelyne had been the first to arrive, landing at St Helena on 6 April. But he was finding it hard to find a suitable place for his observatory on the island. Scrambling up and down the dangerous paths hewn into the steep mountainsides, the stout curate from Chipping Barnet cut an odd figure in the rugged landscape of St Helena. If he slipped, he would 'inevitably be precipitated to the valley', another traveller warned. As Maskelyne climbed up and down the narrow trails, he looked around to find a convenient spot, but nowhere was quite right: the more easily accessible valleys were useless observation points because the surrounding hills blocked the sightlines, while the locations higher in the mountains, Maskelyne complained, were 'almost perpetually covered with Fogs & vapours'. Unsurprisingly he was having the same problems as Edmond Halley, who had observed a transit of Mercury on St

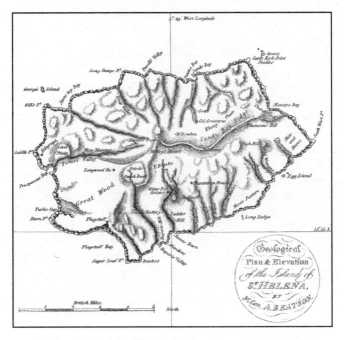

Map of St Helena with Halley's Mount in the middle of the island

Helena in 1677.* Strangely – despite knowing that the island
was 'infested' with clouds – it hadn't stopped astronomers from
recommending St Helena as a site for the Venus observations.
After days of zigzagging through the rough terrain, Maskelyne
settled on a location, 'some distance below Halley's Mount'.

The governor of St Helena assisted as much as he could.
For the construction of the twenty-four- by twelve-foot observatory, he sent workmen and materials – without which,
Maskelyne acknowledged, 'I do not know what we could have
well done', as the observatory 'would certainly never have been
finish'd in time'. Maskelyne oversaw the building and prepared
his instruments to be ready for the transit. In order to set his
clock accurately, he had to work out the exact time of noon
– this he did like all the other astronomers by using the so-called
'equal altitudes' of the sun. Every day – as long as the sky was
clear – he measured the height of the sun above the horizon
a few times before and after noon. From these observations,
he then worked out when the sun reached its highest point
and set his clock accordingly. Despite his diligence, he experienced problems with his ten-foot telescope because 'thro' the
tardiness of the workmen in finishing it', he had not been able
to test it before leaving England. He also continued his observations of the moon and Jupiter's satellites in order to determine
St Helena's precise longitude which was, as he said, of 'utmost
importance'.

Mason and Dixon meanwhile had ignored their orders once
again and made plans to observe the transit from the Cape of
Good Hope. Much delayed – first by contrary winds that had
left them stranded in Portsmouth for several weeks, and then
by the disastrous battle with the French frigate only four days
into their voyage – they were already late when they made a
stopover at the Cape on 27 April 1761. They decided to stay.
It would be impossible to cross 6,000 miles of war-torn waters
during the hurricane season in little more than a month. Not
only that, they had also received news that the French had
taken the British trading port of Bencoolen in the ongoing

_____

* Halley had travelled to St Helena in the late 1670s to prepare a map of the
stars of the southern hemisphere. It was on the remote island that he had
observed a transit of Mercury, which triggered his idea of using the transit
of Venus as a way of calculating the distance between the earth and the sun.

battles for colonial possessions in the East Indies. When Mason and Dixon saw the southern point of the African continent jutting into the ocean like a gnarled finger, they resolved to take their observations there. Protruding high above the sea, the rocks offered unobstructed views into the night sky – just as the Astronomer Royal had directed. On 6 May, with exactly one month to the transit, they sent a letter to England, explaining their reasoning and emphasising that it had been the ship's captain who had made the final decision, smoothly diverting any subsequent blame from themselves. In London, the fellows of the Royal Society remained blissfully ignorant of the fact that their expedition had ended prematurely.

A view of Cape Town with the Table Mountain behind

Sheltered by the imposing Table Mountain, Cape Town was a lively Dutch entrepôt that was frequented by many Europeans sailing towards India, China and the East Indies. As one traveller remarked, it was the most visited 'distant part of the Globe', catering for the growing numbers of colonial sea voyagers by offering temporary accommodation onshore and provisions for their ships. When Mason and Dixon arrived, the bustling town consisted of around 1,000 whitewashed brick houses with thatched roofs, all built along a regular grid of broad roads. Nearby vineyards produced a wide variety of wines and the gardens were planted with European vegetables and exotics so that Mason and Dixon would be able to enjoy cabbages and broccoli, and also fruit such as guavas from the Caribbean. Cape Town was the epitome of Dutch neatness, cleanliness and efficiency. 'Foreigners', another traveller praised, 'find themselves more at home in this port than can be imagined'. Mason and Dixon stayed in one of these houses, for the 'common method of living' in Cape Town was to lodge privately. There was also a large botanic garden with deliciously scented blooms and a shaded oak-tree walk. It was the Dutch East India Company's horticultural storehouse for their empire, filled with plants from South Africa as well as from the Dutch colonies in the East Indies.

The two British astronomers did not have much time to admire the sights. They hired a carriage to transport their instruments and the Dutch governor did all he could to assist the scientific endeavour by providing materials and workmen to help construct an observatory. But Mason and Dixon were quickly becoming impatient. On 6 May, exactly one month before the transit, Mason wrote anxiously to the Royal Society that 'the Dutch are so slow' and did not understand a word of English. As the astronomers tried to explain how to build the observatory by pointing and sketching, they began to despair 'of getting it completed in time'. With the days ticking by, Mason and Dixon successfully entreated their ship's captain to send his carpenters for a few days. Without their help, Mason and Dixon wrote to the Royal Society, they would never have completed the observatory in time. When the carpenters were finished, they had a small tent-like structure – circular with a diameter of six-and-a-half feet, covered with canvas and a conical roof that could be opened 'to any part of the heavens'. They fixed their clock against two pieces of timber which were buried four feet in the ground. They spent the next weeks checking their clock against the height of the sun, so that they could accurately measure Venus's entry and exit. Like Maskelyne in St Helena, Mason and Dixon grew increasingly worried about the weather because it was cloudy 'near all the time'. Unlike the other observers, however, they were doubly nervous because only a successful observation would justify their clear breach of instructions and decision to remain at the Cape of Good Hope.

Tobolsk, the capital of Siberia

In Russia, Chappe was also struggling to get ready. Upon arrival in Tobolsk, he had immediately scoured the area to find a suitable place for his observatory, and had located the perfect spot on a mountain about a mile outside the city. Then, just when he thought all would be well, the entire expedition was jeopardised by the local inhabitants who, as he set up his large telescopes, clock and quadrants, became convinced that he was a magician. They blamed him for the unusual torrents of water that came gushing down the mountainsides and flooded the region as the ice melted. As their houses and fields disappeared, the peasants threatened to murder Chappe. To protect him the local governor provided guards, and the astronomer decided to sleep in his observatory in case the mob 'should attempt to pull it down'.

Shielded from the locals' fury, Chappe hurried to have his observatory finished by 18 May 1761, when he needed to observe a lunar eclipse so that he could calculate his longitude. The eclipse was visible in many parts of the world and other transit astronomers also used it to establish their exact geographical position. In Kajana, in eastern Finland, the Swede Anders Planman was also preparing for the eclipse, but then realised that treetops obstructed his view. The inventive astronomer quickly dragged out three armchairs from the parlour of the local postmaster (with whom he lodged). Piling them on top of each other, and with the bemused postmaster and his wife holding the precarious construction steady, Planman took his telescope and climbed to the top of his almost four metre high armchair-tower to see the earth's shadow veiling the moon. In the same night Mason and Dixon were directing their telescopes towards the disappearing moon from the Cape of Good Hope, while Pingré watched from the deck of his vessel in the Indian Ocean.

With less than three weeks to the transit, Pingré was still a long way from Rodrigues. After his ship had been forced to accompany the damaged French supply vessel, Pingré disembarked on Mauritius on 7 May. The governor promised to help him. Fresh from his experience with Le Gentil, who had waited so long on the island for a passage to India, the governor was clearly becoming accustomed to comforting and assisting desperate French astronomers. He had already found a small ship, he assured Pingré, that could leave the next day.

It was still possible to reach Rodrigues in time – an eight-day journey and no more, one captain had told Pingré. However, eight days later Pingré was still nowhere near. Squalls and high waves had slowed them down at first, only to be replaced by a lull. The ship was not moving at all. Days were ticking by and for him the frenzied race had come to a standstill. On 26 May, Pingré finally saw Rodrigues in the distance – a sight 'that filled me with such satisfaction as I haven't felt since my departure from France', he cried, but there was still no wind. He was now, Pingré believed, in the hands of God and the captain. 'The calm continued on the sea, in the air and in the spirit of M. Thullier [his assistant]', he wrote in his journal, 'but certainly not in mine'. For two more days their ship was as still as if captured in a painting, with Rodrigues as the tantalising background. Then, on 28 May, only nine days before the transit, Pingré finally set foot on the 'desired island'.

There was no town or fort on Rodrigues. The only reason the Compagnie des Indes kept the island was for its large turtle population. Regarded as a remedy against scurvy, the turtles were collected and kept in an enclosure and every two or three months dispatched to Mauritius. The governor of Rodrigues, Pingré snobbishly noted, lived only in a small log cabin made of roughly hewn timber and mud. Pingré and his assistant had to sleep in a shed with a dirt floor beside this governor's 'residence'.

'We had no time to lose', Pingré wrote. He found a location in the north of the island from where to view the transit, but it was too late to build a proper observatory. Instead he placed some big boulders in a circle and constructed a small hut to house the instruments. It was so crudely built that it gave little protection from wind, dust and animals. The instruments had already suffered from the long sea voyage with some 'eaten by rust', Pingré moaned, hectically polishing and greasing them with turtle oil, the only lubricant available. Over the next days, the French astronomer prepared his instruments and observed the movements of Jupiter's satellites at night in order to set the clock – an enterprise that was sabotaged by the rats that chewed through one of the pendulums.

At their many locations across the world, the astronomers all busied themselves with last-minute preparations, united in their endeavour. No matter what their nationality or religion, no matter if they had travelled thousands of miles or had stayed at home, no matter if they had a twenty-five-foot telescope or a handheld tube – they were all striving towards a common goal. In the midst of the Seven Years' War, the astronomers had overcome national boundaries and conflict in the name of science and knowledge. Now, with only a few hours left until the transit, they could do nothing more than hope that the weather would be on their side.

6

Day of Transit, 6 June 1761

As the earth rotated and one location after another emerged from the shadow of the night into daylight, almost 250 astronomers turned their telescopes towards the sky. The light northern summer nights would allow those in Scandinavia and further north in Lapland to see the entire passage of Venus across the sun (which would start after 3 a.m.), but observers in Britain, Germany, France, the Netherlands and Italy would have to be content with the exit because the first hours of the transit occurred in darkness. Tobolsk would see the beginning of the transit just before 7 a.m., the eastern coast of India at 7.30 a.m., and Jakarta just after 9 a.m. South America would experience the entire transit in the dark, as would the North American colonies from Georgia up to Massachusetts.

As the day of the transit arrived, Delisle must have felt proud to see how his call to action had been answered across the world, even if his deteriorating eyesight meant that he himself would be unable to make any precise observations. He had kick-started the project from his small observatory in Paris. He had taken up Edmond Halley's gauntlet and persuaded scientists across Europe to follow his proposal, proving to be a truly inspiring leader. For the past few years, he had calculated, drawn, persuaded and corresponded, using every contact he had in the international world of science to make this moment happen. Now all he could do was to wait and pray that the voyaging astronomers had reached their destinations in time.

The objectives of each expedition were clear. What the astronomers in the scientific societies of Paris, London and Stockholm required for their calculations was the precise location from which each observation was taken and, crucially, accurate timings of Venus's transit as it passed in front of the sun. If astronomers worked according to Halley's method of duration – measuring the entire transit – they had to note four specific points: the moment that Venus touched the sun's outer edge, the so-called external ingress or entry; the time when Venus had fully entered the sun, the so-called internal ingress or entry; the moment when the small black dot began to exit the sun again, the internal egress or exit; and the external egress or exit time, when Venus disappeared completely. For Delisle's method, observers only needed to note either the entry *or* the exit time.

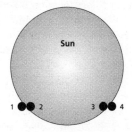

Figure 1: 1. external ingress or entry; 2. internal ingress or entry; 3. internal egress or exit; and 4. external egress or exit

But whatever method they used, no single observation would be useful on its own – scientists needed at least a pair. Depending on their locations, observers would see Venus begin and end her march at slightly different times, or for different durations.

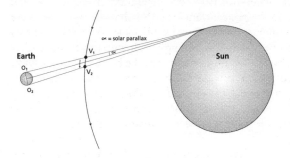

Figure 2: Observer O1 would see Venus entering the sun (V1) first, while observer O2 would see Venus a little later (V2)

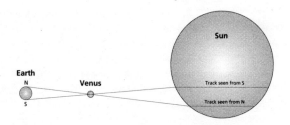

Figure 3: Venus's tracks as seen from observers in the northern and southern hemispheres in 1761

An observer in the south such as Pingré in Rodrigues would see Venus's entry earlier, and on a longer track closer to the centre of the sun, than an observer in the far north, such as Anders Planman in Lapland. The further apart the observers were stationed, the greater the difference between the entry and the exit times, and between the total duration recorded.

What the scientists hoped to extract from this data was the solar parallax: the key to the size of the solar system. A parallax is the difference or shift in position of one object when viewed from two different lines of sight.* The 'solar parallax' (angle α in Figure 2) is measured as the angle (or to be precise half the angle) between these lines drawn from two different locations on opposite sides of the earth to the sun.†

There were different ways to work out the parallax. One was to use the difference of Venus's entry *or* exit times as measured by two different observers (Figure 2) – taking into account the exact geographical position and the differences in local times. The other method was to use the difference in the total duration of the transit as recorded by two observers (Figure 3).

Then, using relatively simple trigonometry, the astromomers

* If you stretch out your arm and look at your thumb against an object in the distance alternately with your left and then your right eye, your thumb seems to jump into a different position. In the transit of Venus, the left and right eyes are the two different observer locations. This would mean astronomers like Pingré in the southern hemisphere (the left eye) saw Venus (the thumb) in a different position against the background of the disc of the sun (the distant object) than another observer would such as Planman in the northern hemisphere (the right eye).

† The solar parallax was measured in arc minutes and seconds. Sixty arc minutes are one degree, and sixty arc seconds are one arc minute.

could begin to calculate the distance between the earth and the sun.* The answer to the enigma of the size of the solar system was within their reach. Or so they thought.

 Le Gentil: Aboard Le Sylphide, *Indian Ocean, south-east of Sri Lanka, just south of the equator, latitude: 5° 44' 10 S, longitude: 89° 35' E*

The sun rose shortly after 6 a.m. into a sky that was scrubbed clean of any clouds. The weather was perfect. In two hours' time Venus would kiss the outer edge of the sun, pushing her dark outline onto the burning disc. Le Gentil would be one of the first to see her, but instead of standing on solid ground in Pondicherry, he was on a moving boat. On the rolling deck he couldn't rely on his pendulum clock, nor would he be able to provide a precise geographical position. But as he thought of all the other astronomers who were also waiting for Venus, he told himself not to be 'idle', and to 'do the best I could'.

He would try to time Venus's entry and exit with a sandglass that ran for thirty seconds – not the most precise timepiece, but better than nothing.† While Le Gentil stared into the sky, one of the sailors would turn the glass again and again, recording the number of rotations made.

Where other astronomers had buried long timbers several feet deep into the ground to make their instruments perfectly stable, Le Gentil had fixed his telescope onto a four-foot timber beam attached to one of the ship's masts. He did not know when Venus would appear. No astronomer knew the exact time of entry – Venus would remain invisible until that very point. This was tiring enough on the eyes in any case, but to determine that precise moment on a pitching ship would be virtually

---

* For example, to work out the distance between the earth and the sun (say 'y'), astronomers could use half the triangle that was formed between the observers O1 and O2 and the sun. Taking half the distance between O1 and O2 – let's call it 'x' – and the angle of the solar parallax 'α', they could calculate the missing side of the triangle, the distance 'y'.

$$y = \frac{x}{\tan \alpha}$$

† The sandglass was meant to run for thirty seconds but Le Gentil recalculated that it in fact ran for thirty-four seconds.

impossible. His best chance was to stare into the glaring sun, waiting to catch a first glimpse of Venus. But whatever Le Gentil tried, he found focusing impossible. Every time the ship rode a wave, he lost sight of the sun. His eyes were already tired even before Venus had shown herself. To save his energy and eyesight, Le Gentil therefore decided to concentrate on Venus's exit only, reckoning that it would be easier to follow the planet's progress across the sun once she had appeared.

After what seemed an eternity, Le Gentil finally saw Venus. As he had expected, he had missed her entry. There was nothing much to do now than to admire her slow path. But once again the skies were playing a cruel game. Over the next hours clouds closed in and Le Gentil became convinced that he had missed his opportunity to see the exit. But then the sky cleared, and at 2.27 p.m., according to his – slightly erratic – timekeeping, he was able to see the small planet seemingly touch 'the edge of the Sun'. At that moment he called out to the sailor who began to turn the sandglass – and by the time he had turned it twenty-eight times, Venus had 'exited' and disappeared. The transit was over but Le Gentil knew that his times could not be used – they were, he admitted, 'far from' precise. In a scientific sense his observations were useless. His voyage had been in vain.

 Chappe d'Auteroche: Tobolsk, Russia

After a fitful night Chappe had risen early. The evening before the transit had been emotionally draining. At sunset, the sky had been clear and as the light slowly disappeared, Chappe had embraced the moment with pathos. The 'perfect stillness of the universe', he wrote in his journal, enveloped him and added to the 'serenity' of his mind. He finally felt to be in equilibrium with the universe – the dangers and exertions of the past months had led him to this feeling of perfect fulfil-ment. But then at ten o'clock in the evening, as he looked dreamily into the night sky, fog began to roll in over the horizon and puffy clouds obscured the stars – vanguards for a thick black cloud that soon pushed in. As the heavens drew their dark curtains and the stars disappeared, Chappe had become so distressed that he was thrown 'into a state of despondency'.

The journey had been 'fruitless', he declared, and he had faced all the dangers 'in vain'.

Though it was the middle of the night, he woke his assistants who slept in the observatory, rushing them out of their warm blankets so that he could be alone in the little hut – not caring where they would spend the night. As he paced up and down the small room, his telescopes seemed to mock him. The cramped observatory felt claustrophobic. Every few minutes Chappe stepped outside, scanning the dark sky in disbelief. He stayed up the entire night, shaking in 'dreadful agitations'.

When the sun rose behind a wall of dark clouds, all Chappe could see was a dim red shadow behind the greyness. Then the wind suddenly picked up and drove the gloomy veil away. With the sky clearing, Chappe perked up, filled 'with a new kind of life'. All nature, he said, 'seemed to rejoice'. His dark mood and thoughts vanished as quickly as the clouds, instead he was joyous and confident. Chappe, who seemed to think and experience only in extremes, felt like a new man.

Shortly afterwards the governor of the region, his family and the archbishop of Tobolsk with some of his churchmen arrived at the top of the mountain, and 'shared my happiness', Chappe wrote. He had pitched a separate tent for them in which they could use one of his telescopes and had doubled his guards – precautionary measures in order not to be disturbed in his own observations. The watchmaker was ordered to take notes and keep an eye on the clock, while the interpreter had to count the minutes and seconds.

When the time of the transit approached, Chappe stood by his telescope looking back and forth, 'between us and the sun a thousand times in a minute', he said. In his excitement he barely noticed a cloud that had settled on one side of the sun. Only when the wind blew it away, Chappe realised that Venus had already touched the sun. Undaunted, he told himself he would concentrate on the moment when Venus fully entered – the internal ingress. As the tiny black circle slowly moved, the excitable Chappe started to tremble. 'I was seized with an universal shivering', he later explained, 'and was obliged to collect all my thoughts, in order not to miss it'. As Venus detached herself from the sun's inner edge, Chappe shouted out so that the watchmaker could write down the time. For the next hours he watched Venus's gradual progress and finally timed her exit. His long journey to Siberia had all proved worthwhile. Convinced of the accuracy of his observations, Chappe was certain that they would be 'useful to posterity, when I had quitted this life'.

 Alexandre-Gui Pingré: Rodrigues, Indian Ocean

Pingré woke in the middle of the night to the sound of heavy rain drumming on the roof of the little shed that he now called his home. Later, when he had left the island, Pingré would calculate that during his stay of 104 days, it had rained on ninety-three . . . not at all what the Académie in Paris had predicted when they had dispatched him.

It was still raining when Pingré rose at 5.a.m., but he was prepared nevertheless. His instruments were cleaned and set up inside the circle of large boulders. His two telescopes were fixed with ropes and rollers on to sturdy timber masts. The clock was in position (after emergency repairs to the pendulum) and the time was set. He stared into the sky as the sun climbed over the horizon at just after 6 a.m., but he couldn't see anything.

Then, a little later, the sun briefly showed itself – with the black dot already present. Venus must have entered just before sunrise and Pingré had missed the beginning of the transit. But over the next two hours, the astronomer occasionally saw Venus through small openings in the thick layers of clouds. At around 8.30 a.m. the sky cleared a little and Pingré with his assistant, as well as the captain and second lieutenant of the ship that had brought them to the island, began to measure the distance between Venus and the inner edge of the sun – although, he said, the results 'cannot be depended upon, to more than one second'. Not only was Pingré short-sighted, he also had difficulties in focusing his telescope because the strong winds 'disordered my instrument'. Seconds stretched into minutes, and minutes into hours. Time crawled. Then, just as he prepared to view the exit of the small planet, a cloud moved in front of the sun. According to his clock, at 12.53 p.m. and eighteen seconds, Pingré thought he could see a 'faint' shape through the clouds which might have been Venus exiting, but he couldn't be sure. Just over a minute later, at 12.54 p.m. and twenty-one seconds, he noted that the transit had 'certainly ended'. His observations, he later admitted to the Royal Society, were taken 'hastily, on account of the clouds'. But instead of despairing, he and his fellow observers celebrated that night, toasting 'the astronomers of all the countries that will have observed Venus tonight'.

In conjunction with observations from the northern hemisphere it might just be enough to provide the key to the distance between the sun and the earth, an optimistic Pingré hoped.

 Mikhail Lomonosov: Lomonosov's house, St Petersburg, Russia

Mikhail Lomonosov's argument with Franz Aepinus had reached a dramatic finale. On 3 June, three days before the transit, the Council of the Academy of Sciences had ordered the German scientist to hand over the keys of the observatory to Lomonosov's protégés. Outraged by the decision, a sulking Aepinus had marched off, announcing that he would not bother to view Venus. Lomonosov's protégés were at the Academy's observatory, while Lomonosov himself had retreated to his own house to conduct the transit observations.

It was the culmination of the row that had begun the previous winter. In May, long after Chappe had left St Petersburg, Lomonosov had insisted that some Russian scientists should join Aepinus in the observatory on the day of the transit. The German, who resented Lomonosov's constant interference, stubbornly refused and wrote a long letter of complaint to the Council of the Academy, presenting rather flimsy arguments as to why he alone should be responsible for the observations. Lomonosov's observers, Aepinus insisted, had 'no knowledge on the topic of observations'. They could also not read Latin, German, English, French or Swedish which meant that they were completely excluded from the international exchange of information about the transit. Without the language skills, Aepinus wrote, Lomonosov's protégés were unable to understand the discussions on how to view Venus, when to time her path, how to make the calculations, which instruments to use and when. And for good measure, he also claimed that they would make too much noise. The observatory was small and they would disturb him, Aepinus argued, so that he would not be able to hear the ticking of his clock (this was important in order to count the seconds while timing the entry and exit times).

Unsurprisingly, Lomonosov had quickly prepared a counterattack, laying out his misgivings in a long letter that rebutted, point by point, Aepinus's arguments, while at the same time presenting his German colleague as being incompetent. Every observatory in the world, Lomonosov argued, invited several co-observers to participate during important events and his protégés were perfectly capable of viewing the transit. They had gazed at the stars while Aepinus was still reading his 'catechism in school'. That Aepinus claimed that he would not be able to hear the clock was just another proof of how unskilled he was – good astronomers, Lomonosov said, could 'count seconds without the need of a clock'.* The whole affair, he maintained, was 'simply madness'. Madness it certainly was, because the result was that Aepinus, who was one of the best astronomers in St Petersburg, decided that assuaging his hurt pride was more important than viewing Venus's rare appearance.

Lomonosov, on the other hand, was not going to miss it. He had woken early and had been sitting at his telescope, desperately focusing on the glaring sun. But there was nothing to see. As his eyes tired, he began to worry. Then after forty minutes of staring straight into the sun, he saw something strange: the clearly defined edge of the disc had become hazy and 'seemed to be disturbed' just at the point where he believed Venus would enter. Quickly Lomonosov turned away from the telescope to rest his 'weary eye' and when he looked again he saw the black dot. Venus was pushing in. For hours Lomonosov then watched her crossing the face of the sun, resting his eyes now and again to be prepared for the exit. Once again the scene was not at all what he and the other astronomers had expected. As Venus readied herself for the exit 'a little pimple'

* Lomonosov was exaggerating – most observers had an assistant counting out the seconds during the transit.

appeared on the inner edge of the sun, reaching out to her and finally touching the smaller planet. Then, as Venus began her departure, the edge of the sun became 'disturbed' again. Although he had left the timing of the entry and exit to his protégés in the observatory, Lomonosov was certain that he himself had made an extraordinary discovery: Venus, he now believed, had an atmosphere just like the earth. The strange haziness at the sun's edge, he noted, had been created by Venus's atmosphere. The black 'pimple' or bulge was a further proof, he explained, because it was the result of the 'refraction of the solar rays in the atmosphere of Venus'. This, he concluded, could even mean that life existed on the planet.*

 Anders Planman: Kajana, Finland

Planman waited in the small observatory that he had built in Kajana. The sun had risen at 2 a.m. and he was expecting to see Venus at around 4 a.m. The instruments were set up, his assistant had the task of counting out the minutes and seconds and Planman had taught the local clergyman how to use one of the telescopes. With his usual diligence and meticulous attention to detail, Planman had prepared for every eventuality. He had determined the longitude after his observation of the lunar eclipse from his tower of armchairs and had erected a simple observatory from roughly hewn planks to protect his instruments from the fierce wind and frost.

Planman woke to a blue sky and was confident of success. Then, just at the moment of Venus's appearance, the air filled with thick smoke. Local farmers had set fire to nearby woodland in order to clear the land. To make matters worse, they had lost control of the blaze and the flames were devouring large swathes of forest. Billowing smoke continued to waft over the little observatory. Planman was furious. Had he only known that local agricultural methods included burning, he told Wargentin later, he would have ordered a 'ban'. The smoke had taken him by surprise and he wasn't sure about Venus's

---

* Lomonosov was not the only one who concluded that there must be an atmosphere on Venus. The German Georg Christoph Silberschlag was the first to publish this theory on 13 June 1761 in the *Magdeburgische Zeitung*.

exact external entry time, but within a few minutes his eyes had adapted and he coolly noted the full ingress.

His assistant was less lucky. The wind shook the clergyman's telescope so much that he couldn't focus and his eyes quickly tired. Assiduous as always, Planman himself had anticipated the latter problem and had long practised observing the skies with either his left or right eye – regularly swapping during the transit so that his right and best eye was rested for Venus's exit just after 10 a.m. Though hampered by the smoke and his assistant's shortcomings, by the end of the transit Planman was pleased with his results. He had noted the ingress (though a little doubtful about the beginning) and the egress.

He had seen, as he proudly wrote to Wargentin, 'the nuptials of Venus' from beginning to end.

 Pehr Wilhelm Wargentin: Stockholm, Sweden

Despite the early hour, the Stockholm observatory was full. Wargentin's publicity campaign had been so successful that it was almost impossible to move in the packed room. Queen Louisa Ulrika and her fifteen-year-old son, Crown Prince Gustav, had arrived in the dark just before 3 a.m. Politicians, noblemen and foreign ambassadors as well as many other spectators were jostling for space. Wargentin had also asked members of the Swedish Academy to assist him and had equipped them with telescopes. Once everybody had squeezed in, it became immediately clear that not all the observers would be able to see the clock – and so it was decided that one of them would call out the minutes and seconds.

As the sun rose, the sky was a polished blue. All the observers stood at their telescopes, waiting for Venus to appear. The weather was perfect, but then Wargentin saw the same phenomenon as Lomonosov did that very moment in St Petersburg: the edge of the sun seemed to be 'boiling', which made it difficult to discern if Venus had already entered. At 3.21 a.m. and thirty-seven seconds, Wargentin spotted a small dent in the sun which became larger and darker – eleven seconds later he was certain that this was Venus, but he did not know which time to note as the exact moment of entry. As Venus moved on to the sun, Wargentin thought that the tiny black dot was now surrounded by a glowing halo but his colleagues couldn't see it. Despite these problems and the noise from the spectators – which made it difficult for the astronomers to hear the minutes and seconds being called out – they noted the external and internal ingress times. They now had almost six hours to enjoy Venus's leisurely march.

Just before 9.30 a.m., everybody concentrated again with 'greatest diligence' for the moment of exit. One of them saw something 'shooting' out of Venus towards the sun, others did not. Once again, Wargentin was certain that he could detect a 'narrow ring' of light, but he was alone in this. As the astronomers timed the final exit, it became clear that their observations didn't correspond. The times noted ranged from 9.47 a.m. and fifty-nine seconds to Wargentin's 9.48 a.m. and nine seconds – for a precise science like astronomy a substantial difference. They did 'not agree so near as was hoped for', a frustrated Wargentin commented.

 Charles Mason and Jeremiah Dixon: Cape of Good Hope, South Africa

Mason and Dixon were nervous. In addition to their worry over the unauthorised stop, the weather at the Cape of Good Hope was appalling. Since their arrival six weeks previously, the sky had been overcast almost every day and night, and they had made few useful observations. But when they had gone to sleep the previous evening, they had been a little more optimistic – the clouds had at last disappeared.

They knew that they would miss the beginning of the transit. Though they were on almost the same longitude as Stockholm, the later sunrise at the Cape would prevent them from seeing Venus's entry. While in Stockholm the sun had illuminated the sky at 3 a.m., Mason and Dixon would have to be patient until dawn, just after 7 a.m. As they waited in the dark observatory, each movement of the clock's hand brought them closer to the precious moment. Tick . . . tock . . . tick . . . tock . . . the regular beating of their pendulum clock was the only sound. Then the sun finally rose but 'in a thick haze', and immediately disappeared behind a dark cloud. Mason and Dixon couldn't see Venus. Time dragged on. Exactly twenty-three minutes after they had first seen the dim light of the sun behind a veil of moisture-laden air, they discovered Venus – briefly – before she and the sun disappeared again. For the next two hours the sun played a game of hide and seek. Just before Venus prepared for her exit, the sun 'was entirely hid by a cloud', but then the sky suddenly cleared. Not moving from their telescopes, Mason and Dixon timed the beginning and end of Venus's exit with nothing disturbing their view. Twenty minutes later the sun once again vanished for the rest of the day, but the first observation of the transit in the southern hemisphere had been a success.

 ## Nevil Maskelyne: St Helena, South Atlantic Ocean

Since his arrival on St Helena two months previously, Maskelyne had stared into an overcast sky, realising with each passing day that his chances of glimpsing Venus's transit were minuscule. For most of the time thick clouds had hung over the mountains of the small island and he was sure it would be the same today. Despite his doubts, Maskelyne had carefully prepared his instruments and clock, adjusting the lenses, tightening screws and bending springs. To his great frustration the clock had stopped the day before the transit, but he had managed to repair it in time.

On 6 June, Maskelyne rose in the dark and set off for the little observatory. In St Helena, as in Europe, the transit would begin at night. The sun rose into a predictably dreary sky but minutes later, and to Maskelyne's great surprise, the clouds parted to present Venus on her golden background. His joy was short-lived as the sun quickly disappeared. An hour later, as Venus prepared for her finale, an excited Maskelyne noticed the sky glowing blue and Venus appeared 'as an intensely black spot upon the sun's body'. As the small circle approached the sun's inner edge, Maskelyne saw what so many other astronomers were observing at the same time: the edges that were about to touch seemed to vibrate and tremble. Like the others, he wasn't sure if this truly was the beginning of the internal exit. The 'degree of exactness' that Halley had hoped for, Maskelyne wrote to the Royal Society, could not be achieved. To make matters worse, the clouds appeared again and prevented him from taking any exit times – 'the principal observation of all'. When the clouds vanished, the sun smiled mockingly into Maskelyne's face, but there was no trace of Venus.

Now it could only be hoped that Mason and Dixon 'have had a more favourable opportunity', Maskelyne said. As he looked into the sky, he was thinking of all the other astronomers who had just completed their observations and was worried that they too might have failed. 'I am afraid', he sighed, 'we must wait till the next transit, in 1769'. Were Halley alive, Maskelyne thought, he would still have been pleased with the international effort. They might not have succeeded – yet – but 'we have done all, that lay in our power'.

 John Winthrop: St John's, Newfoundland

Harvard professor John Winthrop waited silently in the darkness of his tent, at a camp some miles away from St John's in Newfoundland. He would only be able to observe one hour of the transit, just after sunrise, before Venus exited the sun, and he knew he would be the last of all the astronomers across the globe to see it. He knew that Le Gentil had been dispatched to Pondicherry where the French astronomer would already have watched the transit for more than five hours, as would the British observers in Bencoolen, the Russians in St Petersburg and the Swedes in Lapland.

Winthrop was regarded as North America's greatest mathematician – he was a popular teacher (John Adams, the future president of the United States, had been taught by him) and a keen astronomer. Like so many of his colleagues, Winthrop had observed the transit of Mercury but had then focused on Venus as the 'most important phenomenon that the whole compass of astronomy affords us'. The event's rarity also made it an 'exquisite entertainment', Winthrop said. Everybody was talking about the transit – it was *the* 'topic of conversation'.

The only American astronomer who planned to record the

event, Winthrop had sailed from Boston on 9 May, with two pupils in tow and the instruments carefully packed. He had arrived at St John's thirteen days later and immediately looked for a location from where to conduct his observation. Bound by mountains 'towards the Sun-rising', the town offered no suitable place. 'We were obliged to seek farther', Winthrop wrote, and after some 'fruitless' attempts he found a place 'some distance' from St John's. They had no time to construct a proper observatory and instead pitched several tents. The days prior to the transit had been spent in fevered preparations. The clock needed to be set by taking the altitude of the sun at noon, the telescopes had to be assembled and the precise geographical position had to be calculated. Attacked by 'infinite swarms of insects' and with their skin covered in dotted trails of 'venomous stings', the men had worked day and night. But they were ready.

Winthrop had chosen the site of his temporary observatory well. Whereas the entire coastline from the southern tip of Newfoundland up to Halifax in Nova Scotia was buffeted by a heavy storm that morning, the area around St John's, in the north-east of the island, was 'serene and calm' – except for the swarms of mosquitoes. Then, as the hands of Winthrop's clock moved to 4.18 a.m., the sun climbed above the horizon and he saw what he called 'that most agreable Sight, VENUS ON THE SUN'. It would take the planet about thirty minutes before she began her exit and so Winthrop invited the locals who had assembled outside his temporary observatory 'to behold so curious a spectacle' and to peep through the telescope. With the end of the transit quickly approaching, Winthrop returned to his instruments with one of his assistants counting out the minutes and seconds while the other took notes. At 5.05 a.m., according to his clock, the show was over and Winthrop had succeeded in recording both the internal and external exit times.

In memory of the wonderful sight, the locals decided to call the place on which Winthrop had built his observatory 'Venus's Hill'.

*   *   *

In Germany almost forty astronomers and official observers had waited for this 'solemn day'. Many had tossed and turned in their beds before dawn, worried about the weather. In the south, in Nuremberg, one observer had listened to thunder raging all night and in Munich five members of the Bavarian Academy of Sciences stayed up, torn between 'fear and hope'. Clouds hampered many of the German observations. Tobias Mayer, the astronomer who had produced the extensive lunar tables which Maskelyne had used to calculate the longitude on his passage to St Helena, observed parts of the transit from the observatory in Göttingen but couldn't detect much through the greyness. One observer in Bavaria described the sun as being 'devoured by the clouds', while Jesuit astronomer Christian Mayer and his patron the Elector Palatine Karl Theodor despaired about 'the unlucky clouds' over the palace garden in Schwetzingen near Heidelberg.

Others were luckier. In Munich, just as the sun rose with Venus seemingly attached to her, one of the observers joyously cried out: 'Jesus, there she is!' An excited commotion ran through the room and all agreed that it was a 'splendid' sight and a great success. In Halberstadt in Prussia, one observer had 'almost no hope' after a storm raged during the night and the sun rose only as a faint light behind a blanket of clouds. But then, just before Venus's predicted exit, the clouds suddenly dispersed and he was able to note the times. In the east, in Leipzig, the weather had been so disappointing that a frustrated observer resorted to writing a fable about the secret rendezvous which took place between the goddess of love, Venus, and the god of the sun, Apollo, during which he covered his lover with a blanket of clouds to hide her from the view of earthly spectators.

In Leiden, Dutch astronomers only caught 'a glimpse of Venus', while those in Amsterdam also despaired that the dull sky had veiled the transit from them. Wargentin had organised thirty-four Swedish observers at ten locations but many of these too found their observations hampered by clouds. In Russia the astronomer Stepan Rumovsky, who had set out for Nerschinsk near the Chinese border, made it only as far as Selenginsk at Lake Baikal and failed to take the entry times. He managed to time Venus's exit – but the wind shook his telescope so much that he wasn't sure about the accuracy of his observations.

Throughout the British Isles more than thirty observers watched. In London alone there were more than ten official

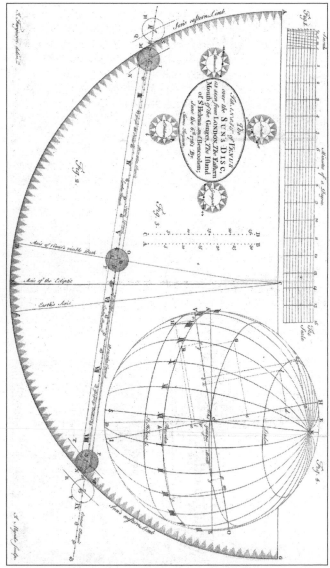

A drawing of the transit of Venus as it was predicted for London, India,
St Helena and Bencoolen on 6 June 1761 – published in James Ferguson's
book on the subject

observers but many 'almost despaired' because of the clouds. Luckily, at half past seven in the morning the sky began to clear, allowing astronomers to note the exact time of the internal and external exits. The tireless Delisle rallied more than forty observers in France. Delisle himself watched from Pingré's observatory at the Abbey of Sainte-Geneviève in Paris, while another astronomer carried his instruments to Louis XV's new Château de Saint-Hubert (twenty miles south-west of Versailles) because the king wanted to observe the transit there. In Italy, astronomers had also been encouraged by Delisle's proposal – which one of them had published in the journal *Novelle Letterarie*. More than twenty Italians in Rome, Bologna, Naples, Turin, Padua, Venice and Parma observed the transit.

Everywhere across the globe, in Europe, Malta, Constantinople, Russia, North America, the East Indies, South Africa, Peking, India, Jakarta and the Philippines astronomers and amateurs watched (or tried to watch) the tiny black dot as it moved across the sun. From the far north of Lapland to the Cape of Good Hope in the southern hemisphere, from Mauritius to Newfoundland, countless observers at more than 130 locations stared simultaneously into the sky in the hope of catching a glimpse of 'Madame Venus'.

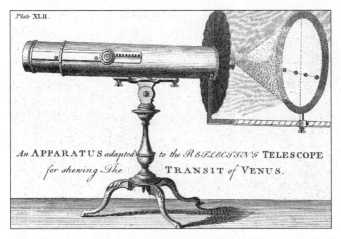

A reflecting telescope with an apparatus to project the image of the transit onto a wall

Many astronomers were joined by their patrons, wealthy merchants and foreign ambassadors, and other curious spectators. In several cities in Germany they used mirrors and lenses to project the image of the sun with the black dot of Venus visible upon it onto walls, so that large groups of onlookers could enjoy the sight. In London, Benjamin Martin's shop in Fleet Street was packed, despite the early hour, for he also projected the image of the transit onto a wall. In Pondicherry, unbeknown to the frustrated Le Gentil, an English officer who was part of the siege of Pondicherry saw the transit there. Together with several others they 'entertained a considerable number of virtuosi' with the extraordinary spectacle – but no one bothered to take the entry and exit times. In Bermuda the guests at a wedding were entertained by Venus when the priest passed around a smoked glass through which to view the transit. Even far north in Lapland the observations in Torneå attracted so much interest that the astronomer there presented the transit on a white screen 'to amuse' the locals. Venus had captured the public imagination.

With the transit over, the expedition astronomers would have to make their way back home or dispatch their results as quickly as possible. Now the astronomers of the European scientific societies faced the difficult task of collecting and compiling all the data gathered across the world. On its own not even the most successful of the observations would be enough. 'The more Observers there are', the *Boston Evening Post* reported shortly afterwards, 'and the more distant their Stations, the more firmly and accurately will the Conclusions be established'. Astronomers waited anxiously for the transit data, hoping that their colleagues abroad would keep their promises and share what they had received.

7

How Far to the Sun?

When Venus exited the face of the sun in the early afternoon on 6 June 1761, Le Gentil had nothing left to do but to stare at the sea in disappointment. He had been the first to leave Europe, travelling thousands of miles through hurricanes and bad weather, as well as surviving disease and deprivation, but had achieved absolutely nothing. Upbeat even in the most adverse situation, he was for once quiet and sullen. He wrote no letters, and the journal that he normally filled with even the most insignificant details remained empty for days after the transit. Nothing could cheer him up.

When his ship briefly docked at Rodrigues twelve days later, Le Gentil was so preoccupied with his misfortune that he missed the opportunity to visit Pingré. Had he left the ship, he would have quickly discovered his colleague from the Académie des Sciences and would have been able to hear first-hand how Pingré had fared. As it was, Le Gentil shut himself away in his cabin, avoiding company and conversation.

His ship continued to Mauritius while Pingré remained behind, still busy with observations to determine the exact geographical position of his observatory. But the Seven Years' War remained an ever-present danger. As Pingré prepared to return home at the end of June, a British ship attacked Rodrigues and confiscated the *Mignonne* and the *Oiseau*, the only two vessels on the island. Rodrigues had no fortifications, and 'half of our weapons did not work', Pingré fumed. The French did not stand a chance against the British. The impetuous astronomer raged and railed at the invaders, angrily waving his passport from the Royal Society – but it was of little use. Rather than recognising the importance of the international collaboration, the British captain instead threatened to hang the French inhabitants.

When the raiding party left a few days later, they hoisted the British flag, took the *Mignonne* and burned the *Oiseau* along with most of her provisions. Pingré was marooned, and given Rodrigues' remoteness, left with little hope of rescue. On the island houses were 'ravaged', poultry, goats and cattle were stolen and slaughtered, fruit taken from the trees and the vegetable gardens pillaged. The eighty people who remained were left with 600 pounds of rice and flour between them and, as Pingré repeatedly complained, were reduced to drinking 'the disgusting beverage of water'.

In mid-July two more British ships landed at Rodrigues. This time Pingré was treated with more 'politeness and humanity' but these captains also refused to free him. They did, however, agree to convey the results of his transit observations to Mauritius, from where the reports could be sent on to Paris and to the Royal Society in England. No matter how furious Pingré was with the treatment he had received, as a scientist, he continued to believe in the international nature of the undertaking. But he remained stranded on Rodrigues, hungry and desperate for a proper drink. By the end of August all the rice was gone, morale was low and the remaining islanders had begun petty fights with each other.

Le Gentil, meanwhile, had arrived in Mauritius, where his old élan was slowly returning. He was lucky, after all: Venus would return in eight years. Under no circumstances would he miss the next transit. It had become the most important reason for his existence. With no wife or children waiting for him in Paris, there was no urgent need to return. Astronomy ruled his life, and astronomy was ruled by Venus during the transit decade. Venus, the brightest of all stars in the sky, the planet named after the goddess of love, was Le Gentil's guiding light. As beautiful and bright as she usually was, Le Gentil only wanted to see her as a small black dot moving across the sun. His passion for her lay in the possibilities of science: the hope that she held the key to the size of the universe. In fact, the tenacious astronomer was so determined that he simply decided to remain in the region rather than sail home to France. He would wait for her – even if it was for eight long years.

Le Gentil only needed an argument to convince the Académie in Paris to continue paying his expenses and salary. He quickly came up with a plan. During his voyage to India he had noticed how imprecise the region's sea charts were. A determination of the exact geographical position of the islands in these seas was essential to navigation and trade, he wrote home, emphasising that it would 'require several years' to complete – noting to himself that this would 'compensate' for his disappointment about the transit as well as allowing him to 'wait for the Transit of Venus of 1769'.

Impatient as always, he did not bother to wait for a response. On 6 September 1761, he informed his superiors in Paris that he had travelled to Madagascar where he intended to make observations 'as useful as possible to geography, navigation and natural history'. He also reported on his transit observations as well as on Pingré's, which had been delivered – as promised – by the British captain who had stopped at Rodrigues. Le Gentil asserted that he had made the 'least bad observation that it is possible to make on a ship at sea', and added – not without a hint of glee – that Pingré's observations had been far from successful. 'He lacks more than half', Le Gentil told the Académie, because Pingré had failed to see the entry times of Venus.

Most of the astronomers who had ventured to distant places quickly dispatched their results, but none seemed to be in a rush to travel home (with the exception of Pingré, who stayed involuntarily). Anders Planman, for example, was reluctant to return to his life of lecturing in Uppsala. He was instead going to Oulu, some 120 miles west of Kajana on the north-eastern coast of the Gulf of Bothnia. Here, he wrote to Wargentin, he intended to 'take the waters' for about a month. It wouldn't cost the Academy anything more than if he remained in Kajana, Planman reassured Wargentin, but the truth was that the journey north and the pressure of the transit observations had exhausted him. 'I got tired of dancing on Venus' wedding', he explained to a friend in Uppsala. To sweeten the idea, Planman promised Wargentin that he would also determine Oulu's longitude while he was there.

Nevil Maskelyne began to conduct experiments on gravity on St Helena, while Charles Mason and Jeremiah Dixon continued to take observations at the Cape of Good Hope, assiduously checking and rechecking the skies to ensure that their longitudinal calculations were precise. By the end of September 1761, almost four months after the transit, Mason and Dixon had finished at the Cape and packed their instruments. At some stage between their arrival in April and their departure on 3 October, they had decided to sail to Britain via St Helena. When they disembarked on the island two weeks later, they met Maskelyne and changed their plans once again. Without further ado, Dixon returned to the Cape with Maskelyne's clock in his luggage to repeat the gravity-measurement tests* there for comparison, while Maskelyne and Mason remained on St Helena. At the end of that year Dixon was back again – by now the astronomers travelled the oceans as easily as if they were taking short carriage rides in London.

Back in the learned societies of Europe scientists began to collate the results of the observations. As detailed letters and long tables of calculations traversed the globe, the astronomers waited in anticipation. Though less thrilling than the expeditions themselves, the subsequent analysis of the huge volume of data was testimony to the single-minded dedication of the astronomers. They were looking at the stars through the prism of Enlightenment thinking, attempting, as many others were, to rationalise and order the natural world.

Only a few days after the transit the first reports were read at the Royal Society in London as well as at the Académie des Sciences in Paris and the Imperial Academy in St Petersburg. Within a few weeks an extraordinary number of letters had been exchanged, travelling across Europe in an intricate web of information. Whoever received such news, copied and forwarded it on to their own correspondents, who in turn dispatched it to their contacts – more circular and looping than linear, but effective nonetheless. Everywhere in Europe, astronomers but also diplomats, merchants and amateur scientists sat at their desks copying out their observations and times – again and again. They folded up their letters, sealed them and gave them to friends or else dispatched them via coach and ship. It was as if an avalanche of knowledge was rolling across Europe.

* They determined the relative force of gravity by measuring the differences in the rate of going of the pendulum clock.

By the beginning of autumn most of the European observations had been exchanged so many times that everybody who was interested knew what the astronomers had seen in Lapland, St Petersburg, Sweden, Britain, France, Germany, Italy and the rest of Europe. Only one week after the transit, a German amateur astronomer and a Dominican scientist in Rome were the first to publish their results, quickly followed by astronomers in Turin and Padua as well as by Mikhail Lomonosov, who in July printed 200 copies of his analysis in Russian and 100 in German. Many observations were published in the journals of the scientific societies such as the Royal Society's *Philosophical Transactions*, the Russian Academy's *Novi Commentarii*, the *Kungl. Vetenskapsakademiens handlingar* of the Swedish Academy and the *Mémoires* of the Paris Académie. Newspapers across the world also reported the results – by September even Boston papers had run articles on the viewings in Newfoundland, London and Pondicherry.

Only the results from the more far-flung expeditions had yet to arrive. Unsurprisingly it took many months for these letters to be received by the learned societies. Maskelyne, for example, sent his results from St Helena in care of his assistant on 29 June, but they took until November to reach the Royal Society. Pingré wrote a letter to the Royal Society in July while still on Rodrigues but by the time it was read to the fellows in London at the end of April 1762, he still had not returned to France. Pingré waited almost three months before a French vessel rescued him and only reached Paris at the end of May 1762.

The Royal Society had to wait even longer to hear from Mason and Dixon. Only in April 1762, when the two astronomers returned to London, did the fellows learn the results of their observations – at which point it also became clear that the best data from the southern hemisphere came from the disobedient pair. With their reputation restored, Mason and Dixon would soon be recommended by the Royal Society to the proprietors of Maryland and Pennsylvania* to end a

---

* In summer 1763, the proprietors of Maryland and Pennsylvania interviewed Mason and Dixon in London before sending them to America.

long-standing border dispute between the two states and neighbouring Delaware and Virginia. From 1763, Mason and Dixon would spend five years in the North American colonies, surveying and mapping the boundary line which was to be known as the 'Mason–Dixon Line' – and many years later became the cultural demarcation between the North and the South (and between the Union and Confederate states during the American Civil War).

Chappe's results took several months to arrive in Paris from Tobolsk. The French astronomer had collapsed immediately after his successful observations of the transit. The physical exertions of his winter journey and the emotional turmoil on the day of the transit had taken their toll. Chappe had vomited blood in the following days and felt an 'overwhelming weakness'. Frail and tired, he had then battled to survive the journey home through swamps and regions infested with Russian rebels. The French consul in Russia wrote to the Académie in Paris that Chappe had made it and that he was well but 'extremely tired, which hindered him from writing himself'. Frustratingly the letter did not mention the transit observations. As was to be expected the Russian Academy was the first to receive the Tobolsk results. The Académie in Paris had to wait almost a year to obtain Chappe's pamphlet on the transit which had been published in St Petersburg and contained all the information they needed.

As the material was collected, the astronomers began their calculations. Poring over the many different descriptions and recorded times, they quickly realised that the observations had not been as successful as they had hoped. Comments like 'doubtful', 'not certain', 'not sure' and 'not accurately' peppered the reports. Once the results were compared it became apparent that many of the observers had encountered similar setbacks.

The problem was that Venus had not moved swiftly on to the face of the sun, but had instead lingered for up to a minute, seemingly glued to the edge, unwilling to embark upon her precious path. At that moment the observers at the governor's house in Madras, for example, had thought that Venus resembled a 'pear' rather than a perfectly circular dot. The astronomers in the observatory in Uppsala had seen Venus both 'in the shape of a drop of water' and as the 'tip of a rapier'. In St Helena, Maskelyne described Venus as 'alternately dilated . . . contracted', and one British observer wrote to the Royal Society that the planet seemed 'to stick to the Sun'. This was what was later called the 'black drop effect'* which made it impossible for observers to determine any exact times of ingress and egress.

What astronomers had expected to see was Venus separating from the sun clearly and quickly. They had assumed, as one German astronomer described it, that the planet would enter and exit the sun 'in the blink of an eye'. In reality Venus's unexpected behaviour made the taking of measurements unreliable – even scientists who had watched side by side had noted different times into their tables. In Uppsala, for example, the difference between the observations recorded by the four astronomers was twenty-two seconds.

But it was not only the black drop effect that had hampered the observations. Many observers had also struggled to determine the exact entry and exit times because the edge of the sun seemed to be 'trembling'. Maskelyne, who only saw Venus briefly but noted she was 'exceedingly ill defined' and Wargentin explained that because of the 'vehement undulations', he could not be accurate about the first external contact – only that 'some part of Venus had occupied the disk of the Sun' at a certain time.

And there was also the strange luminous ring that surrounded Venus on her entry and exit, as noted by Lomonosov and many other astronomers. A German recorded the phenomenon at the observatory of Kloster Berge, a former monastery near Magdeburg, as had other observers in Lapland, Paris, London and Madras. Like Lomonosov, several astronomers concluded therefore that Venus must have an atmosphere similar to the earth's.

---

* The first to describe the phenomenon as 'black drop' was the Swede Anders Johan Lexell who observed the second transit in 1769 in St Petersburg. He described it in Latin in his report on the observation as *'gutta nigra'*. Only in recent years has the true reason for the black drop effect been discovered. It is now believed that it is an optical effect caused by atmospheric conditions on Earth and telescopic diffraction.

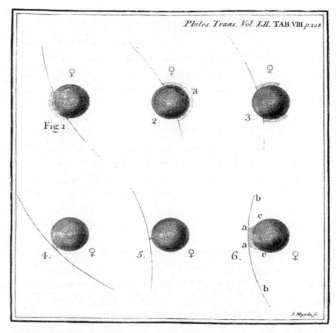

Philos. Trans. Vol. LII. TAB VIII p 228

Figures 1 and 2 show the luminous ring around Venus as she entered the sun. Figures 5 and 6 illustrate how Venus seemed to be glued to the edge of the sun when entering and exiting – the so-called 'black drop effect'.

Nothing about the transit seemed to be straightforward. Planman became so irritated by his own observations that he went over them again and again. Only a week after the transit he told Wargentin that something was wrong with his times. With little else to do in the solitude of Oulu, Planman could not let go. Three weeks later he wrote to Wargentin that either the telescope had failed or that the time of the external exit of Venus had been noted wrongly. When Wargentin sent him the other Swedish observations, Planman compared and calculated back and forth, finally concluding that there was a difference of 'a whole minute'. Under no circumstances would Planman accept that this was his own mistake. Instead he 'suspected' that the blame lay with his assistant, who had counted the minutes and seconds. When Planman went back to his tables and changed the time, Wargentin worried what his foreign colleagues might think about the fluctuating Swedish data.

But Planman was not the only one fiddling with the numbers. A colleague in Uppsala was struggling to reconcile his measurement of Venus's diameter with the results other astronomers had recorded. To make it fit he simply increased Wargentin's figures and his own. In Russia the observer who had gone to Irkutsk admitted that he had 'erred in recording the time', and subsequently altered his results. Pingré amended his data so many times that astronomers in Britain became confused. He had written to the Royal Society from Rodrigues, then corrected those times in a letter from Lisbon on his return journey, and once again from Paris when he realised that the times were still wrong – on the last occasion blaming 'the slowness of the clocks'.

Another issue was the confusion about the exact geographical positions of the various viewing stations. Some observers had used Greenwich as the prime meridian in calculating their longitude, others Paris. This had 'absurd consequences', one British astronomer said, because the exact longitudinal difference between the two had still not been established. Combined with the inaccuracy of Venus's entry and exit times, this meant that more of the essential data for the calculations was flawed.

It also didn't help that the astronomers and scientific societies who had worked together so closely in organising the transit observations and in sharing their results took a more patriotic approach when it came to the calculations – each country racing to be the first to discover the solar parallax and the distance between the earth and the sun.

Delisle, the moving spirit behind the global endeavour, didn't participate much in the calculations. After the transit he slowly withdrew from the astronomical network which for so long had been the focus of his life. Almost eighty, feeble and nearly blind, Delisle had done what he could. With no children of his own, his former pupils like Le Gentil and Jérôme Lalande were now his pride and joy. They were his true 'heirs'. But with Le Gentil on the other side of the globe, it was Lalande who took responsibility for collating the data in France. Lalande had observed the transit in Paris and not joined an expedition only on account of his seasickness, but he enjoyed public recognition and admitted to be an 'oilskin for insults and a sponge for praise'. He seemed to have liked placing himself in the midst of controversy for the sake of the attention. A few years later, for example, he triggered a panic in Paris when he published an article predicting the imminent destruction of the earth through collision with a comet – quietly failing to mention that according to his calculations the event was highly unlikely.

Lalande was so keen to be the first to come up with a figure that he based his initial calculations on Chappe's observations from Tobolsk and those from Sweden alone. He told the Royal Society that his result of the size of the solar parallax was nearly 10" ¼, but then sent a correction two months later when he had received Pingré's observations, reducing it to 9" ½. Converted into miles today, this recalculation equated to a difference of 6,300,000 miles.

Once Pingré returned to Paris in spring 1762, he examined the amassed data and began his own calculations – his result brought the parallax up again to 10"6, while British instrument-maker James Short who had watched the transit in London concluded it was only 8"69. In Sweden, the ever-efficient Wargentin acted as the clearing house for the international data but it was Planman who now became completely obsessed with the parallax calculations, dedicating the next decade of his life to determining the elusive angle. He first suggested that the parallax was between 8" and 8"5, then decided on 8"2 but continued to calculate. The more Planman changed his results and recalculated, the more Wargentin became unsure as to who was right and who was wrong. Pingré, Wargentin wrote to Planman, was a 'strong calculator' and if he was wrong, many others would be also, so 'I am quite at a loss what I shall believe'. Lalande was deeply disappointed. However often the astronomers recalculated, the base data was just too varied – the values for the solar parallax covered such a wide range that it made for a 20,000,000-miles difference, he told a colleague in Berlin.

'There was no reason to be excited about their happy performance', Lalande told a colleague in Berlin. The results were just too imprecise. 'We are collecting all observations we can from all parts of the globe', a British observer wrote to a colleague at the Uppsala observatory in Sweden, 'but I fear there is not any great prospect of determining the problem'. The results did not allow a precise calculation. 'In comparing these observations together', Wargentin wrote to the Royal Society, 'you will perceive, that they do not agree so near as was hoped for'.

What *was* certain was that the parallax was irrefutably smaller than they had previously thought. Jeremiah Horrocks, the only astronomer who had observed the transit in 1639, had believed it to be around 15", which translated to a distance of 54,500,000 miles. Calculations following the 1761 transit ranged from 8"28 to 10"6, much closer to today's accepted value of 8"79. This meant that, according to the 1761 observations, the distance between the sun and the earth was between 77,100,000 and 98,700,000 miles – not very precise, but within the range of today's computations of just under 93,000,000 miles.

Arguments over the calculations became more heated. The Russians were once again at loggerheads. Stepan Rumovsky, who had taken the most easterly observation in Selenginsk* disputed the results and parallax calculations of his colleague Nikita Popov who had travelled to Irkutsk. Over several weeks the two astronomers used every Academy meeting as a forum to air their grievances, with their arguments ping-ponging back and forth. Rumovsky suggested that they should send their data to colleagues in Berlin, Leipzig, and Stockholm for their 'judgment'. Fearing a prolonged battle and humiliation in the European arena, Popov promised to recalculate his results instead, but Rumovsky accused him of seeking 'remedy in delay'. In the meantime Rumovsky went ahead and dispatched Popov's essay to his friends in Europe. After two months of this public fighting their colleagues had enough, and insisted they should stop their 'useless and harmful squabbling'. Although the cantankerous astronomers calmed down, Rumovsky incited more trouble when he presented a letter from Sweden in which Wargentin announced that Popov's Irkutsk observations 'seem almost without use'. Three weeks later Popov gave up and admitted that he and his assistant had noted the time of the internal entry wrongly – they had been out by two minutes.

Other battles had a more nationalistic tone. The Russians, for example, queried whether they should publish the results from the St Petersburg observatory at all because 'the truth of these observations had been disputed by French astronomers'.

---

* The observation in Peking was more easterly, but had not been successful.

James Short in London wrote a forty-six-page article in the *Philosophical Transactions* which refuted Pingré's calculation and stated that the French astronomer had made 'a mistake of one minute' in taking down the time of Venus's internal exit. Minutes and seconds had been added and deducted, Short wrote, and longitudes adjusted to make the calculations fit. In retaliation, Pingré accused Short of using results that were 'uncertain and even altered'. Their battle hinged on the comparison of the only two valid observations made in the southern hemisphere (Maskelyne's failed observation was discounted) – the French one in Rodrigues and the British one at the Cape of Good Hope. Pingré thought Mason and Dixon had determined their longitude inaccurately, while Short insisted that it was the French astronomer who had made the mistake.*

Pingré rejected many of the observations that Short had taken into consideration as unreliable, including those of Dominican scientist Giovanni Battista Audiffredi who had watched the transit in Rome. Offended, Audiffredi in return published another pamphlet and wrote to the secretary of the Académie in Paris to defend his honour and observations, announcing that he would not take part in 'astronomers' disputes'. He compared the observations in Rodrigues with those at the Cape and concluded that Pingré's were useless. He himself came up with a parallax figure of $9"26$ – yet another number.

'These considerations, I say, make me almost despair of the success we look and hope for', one British astronomer wrote to a colleague in Uppsala. Whatever the dispute, they all agreed that 'a definite decision', as Pingré explained, 'will probably not be formed till after the observation of the transit of 1769'. Maskelyne had already predicted these problems in the days following his own failed transit observation. 'We must wait for the better circumstanced Transit of Venus in the year 1769, to settle with precision the Sun's Parallax', he had written to the Royal Society from St Helena.

They had eight years to prepare.

---

* Thomas Hornsby, yet another British astronomer, published his own thirty-page essay in the society's journal, stating that Pingré had decided 'to prefer his own observations to that of Mr Mason' – but then unpatriotically suggested that maybe it had been Mason and Dixon who had made a mistake.

Part 2
Transit 1769

8

A Second Chance

The conditions for the preparation of the second transit were much improved. As Enlightenment ideals swept Europe, its monarchs and ruling powers became increasingly eager to patronise scientific endeavours and expand their own knowledge. Astronomers in Paris, London, Stockholm and elsewhere were hoping for royal support. George III, who was crowned king just before the first transit, was the first British monarch to have studied science as part of his formal education. He was known 'for his love of the sciences' and had been taught physics and chemistry as a boy, holding a particular interest in scientific instruments, astronomy, the quest for longitude, botany and the work of the Royal Society. Spain and France were still ruled by the Bourbon monarchs Carlos III and Louis XV. Natural allies in the Seven Years' War, they also cooperated when it came to arranging a transit expedition to the Spanish-held American West. They too were passionate about science. Carlos III, for example, promoted university research in Spain and Louis XV had been enthusiastic about the stars since childhood and had observed the first transit of Venus, as had Queen Louisa Ulrika of Sweden and her son, Crown Prince Gustav. Christian VII of Denmark became king in 1766, at the age of sixteen, and not only applied to become a member of the Royal Society in London, but also visited the Académie in Paris.

In Russia, the German-born Catherine the Great proved her faith in science when she had herself inoculated against smallpox at a time when the practice was still forbidden in France. She had come to power in 1762 after disposing her husband Peter III, only six months after his accession. Peter had infuriated his people when he had made peace with their enemy Prussia just as Russia was crushing Frederick the Great's army. Several subsequent erratic and unpopular decisions led to an initially bloodless coup – but eight days later Peter III was assassinated under mysterious circumstances, making Catherine the Empress of Russia, but also a usurper and suspected assassin.

The second transit of Venus captivated the ruling families of Europe. Royal coffers were opened to fund at least some of the expeditions and observations. Catherine recruited a small army of astronomers to observe the event, George III commissioned an observatory in the Old Deer Park in Richmond, and even Carlos III, who had not financed any voyages during the first transit, now came to realise the importance of the international venture. From Denmark to Spain, from Russia to Britain, monarchs wholeheartedly supported the astronomers in their endeavour.

The political climate of Europe meanwhile had changed completely since 1761. Peace had returned in spring 1763, with treaties signed between Great Britain, France and Spain, and between Austria and Prussia (Russia and Sweden had withdrawn from the war in 1762). Britain had emerged as the world's greatest colonial power, having gained from France the land east of the Mississippi, Canada and some of the Caribbean sugar-producing islands, as well as Florida from the Spanish. The British retained control in India, although ceded France a few coastal ports (such as Pondicherry) under the condition they remained unfortified trading posts.

War was over, but relations between the ruling powers remained strained. Despite the peace treaty, the old enemies, France and Britain, were cautious of each other. France had suffered humiliating losses and though Britain was now in possession of a huge empire, the cost of the long war and a series of bad harvests had left the Treasury empty. To fill their coffers, the British had introduced the Stamp Act on their American colonies in 1765 – a tax levied on paper and applied to newspapers, licences, books and even decks of cards, which affected almost every colonist. Protest against the controversial Act broke out, with the Americans demanding representation in Parliament in Britain. France was also burdened with heavy debts – the financial effects of the Seven Years' War were felt by all European nations and many were struggling under the strain. But the Seven Years' War had redrawn the map of the world. Unlike in 1761, when the global conflict had prevented astronomers like Le Gentil and Mason and Dixon from reaching their destinations, the prospects for the 1769 transit expeditions seemed greatly improved.

Not only did the political situation make life easier for the scientists, but the astronomical conditions were also better for the second transit. The advantage of the 1769 transit was that the difference in its duration as seen from two places in the northern and the southern hemispheres would be greater than in 1761 – this would enable the astronomers to make more precise calculations of the solar parallax. After the successes and failures of the first transit, the scientific community came together once again. They also knew that for them there would be no second chance, because after 3 June 1769, Venus would not traverse the sun for another 105 years. Europe's astronomers were determined to profit from the lessons learned in 1761. 'The knowledge of the errors' from the first transit, one British astronomer said, would help them 'to put in practice every method of solving this problem'.

Ambitious as always, French astronomer Jérôme Lalande now took the baton from the frail Delisle and published a mappemonde in 1764 that presented once again the best locations from where to view the transit.* By comparing the results of the 1761 observations, one British astronomer concluded that all the observers who had failed to take precise times had done so because of 'vapours' which occurred when the sun was too low on the horizon. In Uppsala, for example, where the transit had begun shortly after sunrise, the four observers had not been able to agree on the exact moment of entry – their times had been twenty-two seconds apart. Six hours later, when the sun had been higher at 9.30 a.m., the differences between their observed exit times had been only six seconds. As such astronomers were now advised to seek out viewing

---

* Lalande also went to London, in 1763, to discuss the transit. He attended several meetings at the Royal Society.

locations where the transit would occur in the middle of the day when the sun was at its highest.

Returning to their tables, the astronomers calculated that the second transit would be visible in its entirety from the South Pacific, the eastern parts of Asia and the Russian empire, and the northern parts of North America as well as the American West. To their great frustration Europe would only witness the very beginning of the transit just before sunset while the rest of Venus's march across the sun would happen in the darkness of the night – the only exception being the far north where the light summer nights would allow astronomers to follow Venus until the early morning hours.

According to the predictions, the longest durations of this transit would be in northern Lapland and at the Arctic Circle, while the shortest would be in the South Pacific, or as it was then called the 'South Sea' – making them the perfect pair on which to base the calculations. The British suggested that astronomers should travel to Vardø at the most north-eastern tip of Norway (then also called Wardhus or Wardoe), Torneå in Lapland at the northern edge of the Gulf of Bothnia, or 'any other place near the north cape'. The northern observations should be relatively straightforward – the advantage of Torneå, for example, was that its longitude was already known – but finding counterparts for it in the southern hemisphere would be more problematic.

The best place for viewing the shortest transit would be in the vast emptiness of the South Sea, somewhere around 55°S latitude and 155°W longitude.* The difference in the length of the transit duration between Torneå and the South Sea would be an amazing twenty-four minutes and thirty-three seconds – more than double the length between any other two viewing stations in 1761. Though this would make the calculations more accurate, it posed its own problems. Not only was the South Sea on the other side of the globe, but there were also no established trade routes nearby, which meant that astronomers could not simply hitch a ride on a merchant vessel. Worse still, no one in fact knew if there was land there at all. The few

---

* This would be roughly between Australia and South America, along the latitude of Cape Horn – and, as we know today, nowhere near any land.

intrepid explorers who had ventured to the region had returned with accounts of its unimaginable vastness – today we know that the Pacific Ocean covers one-third of our planet's surface.

Thomas Hornsby, a British astronomer and professor at Oxford University who had been deeply involved in the fraught parallax calculations, scoured the Bodleian Library to find historical accounts of early voyages into the South Sea. The results were not entirely satisfactory because only a handful of explorers had ever gone there. At the end of the sixteenth century, for example, a Portuguese navigator called Pedro Fernandes de Queirós had discovered a group of islands at 15°S latitude. The only problem was that these islands lay too far west for the transit to be viewed in its entirety. The Spanish had discovered the Solomon Islands in 1568, but as hard as Hornsby searched, he could find no record of their precise geographical position. It was questionable 'whether there be any such islands', he informed his colleagues at the Royal Society. There were others around 170°W longitude, discovered by Dutch explorer Abel Janszoon Tasman in the seventeenth century. These again were not exactly where the astronomers needed to be, but as the difference in viewing the transit there from seeing it in Torneå would be at least twenty minutes, it was thought to be good enough. If it proved impossible to dispatch an expedition to the South Sea, Hornsby said, the next best place would be Mexico or other parts of the American West where the entire transit would also be visible and where Venus's march would still be much shorter than in Torneå.

Meanwhile the French were busy making their own calculations. Alexandre-Gui Pingré, who had settled back into his comfortable life in Paris after the privations of Rodrigues, presented similar ideas to the French Académie des Sciences, emphasising that 'we have advantages offered to us in 1769 that we lacked in 1761'. Like Hornsby, Pingré concluded that an observation in Lapland should be paired up with a location in the South Sea, although he thought that the sun would be too low in Torneå and therefore suggested an expedition even further north. Like the British, Pingré also turned to the travel accounts of early explorers in order to locate a suitable viewing station in the southern hemisphere. At the end of his search, he provided his colleagues with a list of possible destinations, recommending the Marquesas Islands (which are today part of French Polynesia). The difference in transit duration as viewed there would be twenty-six minutes and forty seconds, Pingré claimed, and they knew roughly where to find the islands. Even better the natives were, according to the historical travel accounts, 'of sweet character'.

An expedition to the South Sea would be dangerous and expensive, but, the British stressed, 'posterity must reflect with infinite regret' if they failed to mount one. The advantages to astronomy and navigation were obvious, but there were also the economic implications for a growing empire. 'A commercial nation', it was suggested, might 'make a settlement in the great Pacific Ocean'.

As organised as the scientific societies were, one astronomer was already on his way. Le Gentil was once again the first to set out, this time from Mauritius. In the years after the first transit he had travelled the Indian Ocean, using the island as his base. He had also sailed to Madagascar several times to improve the sea charts of the region by establishing the exact geographical positions of all the important sites along the coast of the island, as well as investigating wind patterns at sea. His map, he proudly announced, provided 'much more safety for navigation than any other before'.

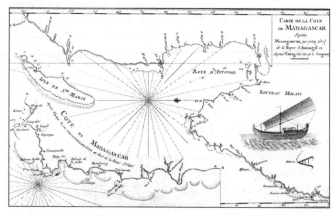

One of Le Gentil's maps, which he produced while waiting for the second transit

By 1765 Le Gentil said that 'it was time to think about the second transit of Venus'. Looking at his astronomical tables and books, he calculated the times for the transit and decided that Manila in the Philippines (then in Spanish hands) would be the best place from which to make his observations. As Mauritius was a stopover on the French trade route, Le Gentil planned to catch a Compagnie des Indes ship that was sailing to China. From there, he had been told, it would be easy to find a vessel to take him on to his destination. Though this was a somewhat roundabout voyage, with four years to spare, Le Gentil wouldn't need to worry. He had plenty of time.

In January 1766, he wrote to the president of the Académie in Paris, requesting letters of recommendation from the Spanish ambassador in France for the governor of Manila. For once serendipity played Le Gentil a good hand. As he was dispatching his letters, a Spanish vessel arrived in Mauritius that was sailing directly to Manila. The ship's first officer turned out to be an old acquaintance from Paris, and the captain was quickly persuaded to assist the French astronomer. On 1 May 1766, after some delays caused by bureaucratic formalities and a hurricane, Le Gentil left Mauritius for the Philippines.

France was once again ahead of Britain. But not for long, the British hoped, because on 5 June 1766, as Le Gentil crossed the Indian Ocean, the fellows of the Royal Society met at their headquarters in London and resolved unanimously 'that one or more astronomical observers be now sought out and engaged to observe the next transit of Venus in 1769'. It was time to consider the practicalities. For the last transit they had only had one year to prepare, but this was their last chance to see Venus crossing the sun. Some of the astronomers would have to travel far away from the existing trade routes on specially chartered vessels which would require more preparations and greater funding.

Slowly the Académie in Paris and the Royal Society in London were turning from their calculations and predictions to the more pragmatic aspects of the observations: how many astronomers should go, whom should they dispatch, which instruments should they take, how would the observers get to their destinations and who would pay for it?

Only Chappe d'Auteroche, who had returned to Paris from Russia in August 1762, refrained from becoming too involved in the intricacies of predictions and preparations. He was in no hurry to leave his position at the Royal Observatory in Paris. After the adventures in Siberia, his life had slowed to a more pedestrian pace. He watched less exciting celestial events such as solar and lunar eclipses, tested new telescopes, and sometimes observed the sky together with his colleague Pingré. In 1764 he conducted some short sea trials off the coast of Brest to investigate the precision of French clockmaker Ferdinand Berthoud's timepieces in order to establish longitude on board ship (the alternative to using Maskelyne's lunar method). Chappe's most pressing occupation during the years between the transits, though, was the completion of a monumental three-volume account of his Siberian expedition, *Voyage en Sibérie*. Chappe had been working tirelessly on this publication which was not just an account of his transit observation but a comprehensive treatise on Russia, including a survey of its climate, minerals, flora and fauna as well as comments on social customs, arts and sciences. Elegant engravings and maps would illustrate the folio-sized volumes, and it was to be his legacy beyond the world of astronomy.

In London the Royal Society was organising an expedition to California, a region controlled by the Spanish. Though not as southerly as the suggested locations in the South Sea, California was nonetheless important. The entire transit would be visible there and occur in the middle of the day. And though the difference in duration would not be as great as it would be in the South Sea, it would still be seventeen minutes – enough, they hoped, for precise parallax calculations. Even better, if the Spanish agreed to cooperate, California could be reached along the existing Spanish trade route from Cadiz to Mexico.

The problem was that relations between the British and the Spanish remained strained, to say the least. During the Seven Years' War Protestant Britain had declared war on Catholic Spain. Allied with France, Spain had suffered great military losses and had ceded Florida to the British after the conflict ended. In early summer 1766, the president of the Royal Society wrote to the Spanish ambassador in London requesting permission to send an astronomer to California. To placate the Spanish, who were always suspicious of non-Catholics, the Royal Society chose a Jesuit astronomer who worked in Italy but was also a fellow of the British society.* In August 1766, Carlos III gave his permission on condition that the observer should travel on a Spanish vessel and be accompanied by 'some spanyards'. Everything seemed to be arranged but then, only a few months later, in March 1767, the Royal Society received the devastating news that Carlos III had given orders to suppress the Society of Jesus – and that all Jesuits should be expelled from Spanish dominions. With one stroke the entire expedition was rendered impossible. Another astronomer would have to be found, and then the Royal Society would have to reapply to the Spanish for permission to enter California.

On 15 May 1767, the Royal Society wrote again to the Spanish ambassador, asking if they could dispatch two British astronomers instead, underlining the importance of this 'unique observation'. The Spanish, however, had lost interest. The Council of the Indies in Madrid, which was effectively the colonial government of the Spanish territories, had no intention of allowing the English into their dominions, believing the expedition was a pretext for espionage. With the English already controlling the eastern seaboard of the North American continent, the Spanish feared for their trade in the West. Under no circumstances would they permit an Englishman to set foot on their North American territory – not even in the name of science. 'No permission will be given to the English astronomers', they wrote, 'such a leave will never be granted'.

The Royal Society's request was now also regarded as an insult to Spanish honour – they had, after all, perfectly competent scientists of their own. The Royal Society, so the Spanish wrote, should be told that 'good astronomers and mathematicians are not lacking in Spain'. Spanish patriotism was awakened. Though they had not shown much interest in the transits so far, they now resolved to mount an expedition of their own. 'His Majesty', the letter continued, 'will give the necessary orders'.

The British were not the only nation to experience setbacks.

* This was Jesuit priest Roger Joseph Boscovich.

Le Gentil had again run into problems. Having been smiled upon by good fortune when he had found a ship to Manila, his luck had not held. During the journey to Manila, he had encountered a 'horrendous storm' which had pushed the vessel precariously close to the coast of the Philippines. Convinced that they would be shipwrecked, the captain had suggested prayers and 'alms'. Faced with death, Le Gentil decided to skip the prayers and instead vowed that if he survived, he would calculate Manila's longitude – the vow of a true but obsessed scientist. Whether it was the captain's prayers or Le Gentil's vow, they made it to Manila alive. He was, as Le Gentil wrote to a friend after the ordeal, 'determined to suffer anything' for the success of his observations. Then in mid-July 1767, just as the British learned of Spain's refusal to let them observe the transit in California, Le Gentil received a letter from Paris that would send him on yet another voyage. The Académie instructed him to leave the Philippines because, according to Pingré's calculations, Le Gentil had gone 'too far east'. They advised that Le Gentil should proceed to Pondicherry instead.

Where other scientists might have been furious or frustrated, Le Gentil remained stoical and considered calmly his options. He decided, yet again, to continue his journey. Only the weather made him hesitate, because out of the ninety days that he had spent in the Philippines, the sky had been overcast on only three. The climate in Manila seemed to be perfect for the transit observation, but, so he tried to reassure himself, it might still be cloudy in Manila on the crucial day in June 1769.

He left with the words of Virgil on his mind: 'Flee this cruel land, flee this flinty shore'. Fired by his usual optimism and unwavering belief that it was his destiny to see Venus, Le Gentil packed his bags and set sail.

9

Russia Enters the Race

As France and Britain prepared for the second transit, Russia hurried to join the race. By order of Catherine the Great, this time the Russians would take charge of their own expeditions. The Empress made it her personal task to get the project under way. In spring 1767, she wrote a letter to Count Vladimir Orlov, director of the Imperial Academy of Sciences in St Petersburg (and brother of her lover Count Grigory Orlov), ordering that preparations for the transit observations should be carried out with 'the utmost care'. Her aim was to ensure that Russia's contributions would be at least as significant as those of France, Britain and the other European countries. When Catherine had seized power in 1762, she had been disappointed to learn that the only accurate observations made of the first transit on Russian soil had been taken by Chappe d'Auteroche, a Frenchman. The 1769 transit, she decided, would be an opportunity to prove to the international community of philosophers and scientists that Russia was not the uncultivated backwater it was generally believed to be.

In the first half of the eighteenth century, Peter the Great had attempted to reform Russia so that it bore 'some resemblance to other European states', as one Englishman said, but the country was still in danger of 'relapse'. Maps of Catherine's empire were divided into 'Russia in Asia' and 'Russia in Europe', and many travellers regarded it as an 'oriental empire'. Catherine was keen to change this perception and wholeheartedly embraced science as a way of demonstrating her belief in progress and rational thought. But it wasn't going to be easy. The country's sheer scale made it almost ungovernable.

Distances were so vast that an order sent from St Petersburg to Kamchatka in the Russian far east took eighteen months to arrive. 'Half of Russia may be destroyed', one traveller said, 'and the other half know nothing about the matter'. In addition, the expense of the Seven Years' War had left Russia's finances in a precarious situation: taxes had remained un-collected, soldiers had not been paid and the administration was in general disorder. But none of this would curb Catherine's ambition.

Within a month of her *coup d'état*, she had made contact with Voltaire – the Zeus of the Enlightenment – or, as Goethe later said, the man who 'governed the whole civilised world'. Catherine knew exactly what she was doing: Voltaire was so influential in eighteenth-century Europe that his friendship would cast her as an enlightened monarch. In her second letter to him she offered to fund a translation of d'Alembert and Diderot's groundbreaking *Encyclopédie*, the greatest collection of knowledge in the eighteenth century. Her determined approach worked. 'It was certainly Voltaire who brought me into fashion', Catherine later admitted to a friend. He, for his part, was delighted to be credited with inspiring the transform-ation of an entire empire. The murder of Catherine's husband was just a 'trifle', Voltaire said, 'family matters in which I do not interfere'.

Presenting herself in the role of munificent patron of scien-tific education, she bought Diderot's library when he ran out of money – allowing him to keep it and even paying him an annual stipend to look after it. Voltaire thoroughly approved, writing to her that 'all the men of letters in Europe must be at your feet'. Catherine's enthusiasm was boundless and she seemed tireless, rising early every morning to fill much of her day with reading and writing. The empire would be different under her rule, she insisted, because 'Russia is a European power'. At the same time as she was instructing the Academy in St Petersburg to organise the transit expeditions, for example, she was also trying to reform Russian law based on her own readings of the French philosopher Montesquieu's analysis of the relationship between the citizens and the state. Her instructions to the Legislative Commission revealed similar intentions to those behind her championing of the transit observations: both were proclamations that Russia was not rooted in Asian despotism but an enlightened European state.

Knowing Catherine's fascination with science, the Imperial Academy had honoured her coronation in 1762 by publishing a special edition of Stepan Rumovsky's viewing of the first transit in Selenginsk (although it had not been a successful one). At the same time Franz Aepinus, who had been tutor to Catherine and to her son Paul, brought the second transit to her attention. Since his battles with Mikhail Lomonosov, Aepinus had increasingly withdrawn from the Academy and instead concentrated on his career at Court. Yet, as an astronomer, he was well aware of the importance of the Russian locations to the forthcoming transit.

The transit observations became an integral part of Catherine's endeavour to align her country with Europe and the spirit of the Enlightenment. Catherine was convinced that Russia needed to play a leading role in the 1769 transit projects. Unlike other European monarchs, though, she did not wait for the astronomers to approach her, but took the initiative herself. She remained closely involved in even the smallest details during the preparations for the expeditions. Just before she set out on a tour of the provinces along the Volga in spring 1767, she requested information about the 'most advantageously situated places of the empire' and asked if workmen had already been dispatched to erect observatories there. Anticipating the same problems as the Academy had experienced in 1761 with the shortage of suitable Russian astronomers, she suggested they should train some naval officers in making observations.

On 27 March, her instructions were presented to the excited members of the Academy who had discussed the subject during meetings over the previous years. Lomonosov had been the driving force behind these early attempts but he had died, aged fifty-three, in early 1765, and without his motivating energy, interest in the project had dwindled.* Now, with

---

* In October 1766, another astronomer had renewed the efforts and told the Academy that the 'training of some younger scholars in the elements of astronomy' was essential if Russia wanted to take part. But again, nothing was achieved.

Catherine herself taking an interest, everything changed. Once her letter had been read, the scientists discussed how best to reply. After several heated meetings, the members of the Academy finished a long letter, which provided all the information Catherine had asked for. Several expeditions to different locations in the vast Russian empire would be necessary, the academicians explained. The entire transit would only be visible in the very north of Russia, but the exit could be viewed throughout the empire.

Like the French and British, the Russians also emphasised that their astronomers would have to establish the longitude of the viewing stations – an important by-product of the expeditions because only twenty locations in Russia had so far been determined accurately. They suggested that at least four expeditions should be mounted: two to the north – to Kola in Russian Lapland and the Solovetsky Islands in the White Sea; to the south, they suggested an observation at Astrakhan at the Caspian Sea and at Guryev at the mouth of the Ural river (today's Atyrau in Kazakhstan).

With Lomonosov dead, the thirty-three-year-old Stepan Rumovsky – who had observed the first transit in Selenginsk and now headed the St Petersburg observatory – was put in charge. As each expedition would need to be manned by at least two skilled observers, Rumovsky agreed to train them during the next year 'with all the possible effort', and tried to recruit twelve young naval officers who were 'eager' and knew the 'basics of mathematics'.

With lessons learned from the first transit, the Russian decided that each expedition should take exactly the same set of instruments, including quadrants as well as telescopes of varying lengths, ranging from three, six, fifteen and eighteen feet. Orders were placed with the best craftsmen in Paris and London. The observatories, they decided, should not be prefabricated because they would have to be designed to suit the particular conditions of each location.

Over the next few weeks, the members of the Imperial Academy wrote to their foreign colleagues and tried to organise the astronomical training of the naval officers. Letters arrived in response from Germany and France, congratulating the Russians on their plans and offering assistance. But no matter

Edmond Halley (1656–1742), the astronomer whose idea it was to use the transit of Venus as a way of measuring the distance between the sun and the earth.

Joseph-Nicolas Delisle (1688–1768) took up Halley's challenge and encouraged his international colleagues to contribute to the transit project. He did not live to see the second transit.

Alexandre-Gui Pingré (1711–96) travelled on behalf of the French Académie des Sciences to the island of Rodrigues in the Indian Ocean for the first transit and observed the second in Haiti.

British astronomer Nevil Maskelyne (1732–1811) went to the remote island of St Helena in 1761, and became the driving force of the global endeavour for the expeditions in 1769.

Russian scientist Mikhail Lomonosov (1711–65) fought over the transit observations with his German colleague Franz Aepinus, and observed the first transit in St Petersburg.

French astronomer Jean-Baptiste Chappe d'Auteroche (1722–69) observed
the first transit in Siberia and the second in Baja California.

The Swedish scientist Pehr Wilhelm Wargentin (1717–83) was the mastermind behind the Swedish observations in 1761 and 1769.

Catherine the Great (1729–96) ordered eight expeditions to all corners of the Russian empire to view the second transit in 1769.

James Cook (1728–79) sailed to Tahiti with Maskelyne's former assistant Charles Green to observe the 1769 transit in the southern hemisphere. He was not only the captain of the *Endeavour* but was also paid by the Royal Society as an observer.

Jesuit priest Maximilian Hell (1720–92) – here dressed in Sami clothes – observed the first transit in Vienna and travelled to Vardø in Norway for the second. Note the drawing of the observatory in the background.

During his stay in London, Benjamin Franklin (1706–90) helped the academicians to organise the observations of the second transit in the American colonies.

American astronomer David Rittenhouse (1732–96) built his own instruments and viewed the second transit from his observatory in Norriton, near Philadelphia.

The Royal Observatory, Paris: the terrace on the garden side, with men experimenting with astronomical and other scientific instruments.

This is one of five clocks made for the Royal Society by John Shelton for the transit expeditions. They were used by the astronomers to ascertain the exact times that Venus entered and exited the sun, and to determine their geographical position.

This reflecting telescope was made by the famous London instrument-maker James Short and is one of the many telescopes that were ordered by astronomers across the globe for the transit observations.

This twelve-inch portable astronomical quadrant was made by London instrument-maker John Bird for one of the many expeditions that were organised by the Royal Society. The quadrant that the Tahitians stole from Fort Venus was exactly the same.

how efficient the scientists at the Academy were, things were not moving fast enough for the Empress. As Catherine travelled along the Volga that spring and early summer, she could not let go of the transit projects and frequently pestered Vladimir Orlov, asking if he had received 'further reports from the Academy'. She requested more information about the suggested locations to ensure that the astronomers would be safe there, and upped the effort, ordering more observers to be sent to the north because it was often 'foggy'. If each viewing was backed up by a second one, taken some distance away, the chances of successfully recording the transit would be much higher, she insisted. At the same time she 'immediately granted' a 6,000-ruble contribution towards the purchase of the best instruments.

When Catherine returned to Moscow from her Volga tour in June, she ordered Aepinus to enquire about the progress of the transit enterprise. By the end of July she was growing increasingly impatient. To avoid misunderstandings and to speed up matters, she sent Orlov to St Petersburg with new instructions.* Aepinus had questioned timber merchants from Kola to determine the most suitable locations there, and Catherine now ordered that all the materials necessary to build the observatories should be 'dispatched' to Kola. Little cabins in which the astronomers could live and work must be constructed in advance, she said.†

In the meantime, Rumovsky was having problems finding suitable astronomers. Some of the naval officers had no inclinations to risk their lives for such an enterprise, while other potential recruits, the Academy learned, were known for their 'low morals'. As a result the scientists had little hope that the expeditions could be manned with Russian astronomers only. Catherine's solution was to lure European scientists into her service with promises of double wages and prestige. Nothing

---

* Catherine couldn't go herself because she attended the opening of the Legislative Commission on 30 July in Moscow, which discussed her 'Great Instruction' – 500 articles, illustrating her vision of how Russia should be governed based on her reading of French and English philosophers.

† The small houses were built before the astronomers arrived, and the observatories – designed according to the Academy's instructions – were later placed on top of the buildings.

was going to prevent her from promoting the Russian empire as the seat of Enlightenment thinking.

During autumn 1767 the academicians were busy with the preparations and in October, they received news from one of London's best instrument-makers, James Short, acknowledging the Russian order. Everybody was relieved as they had 'already began to doubt' if their wish for standardised equipment for each expedition could be fulfilled. Short's assurances, Rumovsky said, 'delivered us from a very disagreeable situation', for otherwise he had no idea how he would have explained the lack of instruments to his impatient empress. In his reply to Short and the Royal Society, Rumovsky then laid out their plans for the expeditions as well as sending a copy of Catherine's initial request for the transit observations. Once read out at the Royal Society in London, Rumovsky knew, the information would be disseminated and the whole of Europe would hear about Russia's scientific endeavour. In the same letter he also proudly reported that Catherine had doubled the Russian effort from four to eight expeditions because of the 'unpredictability' of the weather.

By the end of October 1767, with all the instruments ordered, several astronomers from Germany and Switzerland engaged and the destinations finalised, everything seemed to be on track for the Russian transit observations. The Academy decided to write a report to Catherine summarising their progress. They would send four teams to the north, two to the east and two to the south – the expedition leaders were four Russians, three Germans and an astronomer from Geneva.

But Catherine was not satisfied. In keeping with her Enlightenment ideals, she expanded the remit of the expeditions, ordering that the observers journeying south and east be accompanied by other scientists. She now decided to turn the transit expeditions into comprehensive scientific projects – not only were the astronomers to observe Venus and determine geographical positions of towns and landmarks along the way, but other teams travelling with them would make natural history collections, as well as compiling reports on agricultural and mining opportunities. Each expedition was to have an astronomical leader, two assistants, soldiers and interpreters, as well as a clockmaker, a hunter, a taxidermist and a miner, besides other scientists in the fields of botany and zoology.*

At around the same time as the Academy sent Catherine its report, the Empress also received a letter from Voltaire, who wrote that the 'whole of Europe is looking to the great example of tolerance which the Empress of Russia is setting to the world' – five years into her reign the changes she had introduced were finally being noticed elsewhere in Europe. By the end of the year even American newspapers were reporting on Catherine's eight expeditions across the Russian empire, mentioning that this was done 'at the expense of the Empress'.†

Over the next few months the Russian preparations continued – the only problem was that neither the astronomers from Geneva, nor the German astronomer Georg Moritz Lowitz from Göttingen, who had offered to travel to Yakutsk, had as yet arrived in St Petersburg. In late February 1768, the academicians resolved that they couldn't wait for Lowitz any longer and that instead they had to send a Russian team to the far east of the empire. Considering the size of the Russian empire these delays seriously compromised the expeditions' chances of success.

It was a wise decision because it took several more weeks before Lowitz arrived from Göttingen. The widowed astronomer also brought his eleven-year-old son with whom he intended to travel across the vastness of the empire to his transit observation station. One and a half months later, at the end of May, finally the last group of foreigners arrived in St Petersburg from Switzerland.‡ One year before the transit the crews for the expeditions were complete. Tempted by

---

* One of the German astronomers, Georg Moritz Lowitz, was also a surveyor and was given a special assignment after the Venus observation at the Caspian Sea. He was asked to investigate the viability of a canal to connect the Volga and Don in order to improve trade in these distant regions.

† Later a Russian newspaper insisted that it had been Catherine's efforts which had encouraged other astronomers across Europe to observe the transit. Her engagement 'caused' others to join the project.

‡ Jacques André Mallet and his assistant Jean-Louis Pictet from Geneva.

double salaries, adventurous expeditions and the prospect of glory, the foreigners had travelled thousands of miles already. At the end of May, Catherine granted an additional 10,000 rubles. All they needed now were their instructions and equipment.

What Catherine and the Russian astronomers did not know was that overworked James Short had become seriously ill – perhaps unsurprisingly – after being inundated with requests from astronomers in Britain and across Europe. Everybody wanted the best telescopes and quadrants to ensure the success of their own observations.

Over the past century and a half man's ability to view the solar system had improved dramatically. Until the invention of the telescope in the seventeenth century, the naked eye had been the only way to look at the stars. Early refracting telescopes (which used a lens to form an image) had then been replaced by reflecting telescopes (in which a combination of curved mirrors reflected the light and formed an image while reducing distortion). The latest innovation was the so-called 'achromatic' lens which was a compound lens made of crown and flint glass. This combination of two types of glass compensated the different dispersions. Previously the dispersion of light had been reduced by constructing very long (and unwieldy) telescopes, but achromatic lenses now gave five-foot telescopes the same sharpness as the traditional twenty-foot ones. It was, one German astronomer said, 'a very happy invention which England alone was capable of production!'

During the first transit, few astronomers had used telescopes with achromatic lenses, but with the black drop effect and uncertainties about the timing of Venus's entry and exit, the quality of the observers' telescopes was crucial. The order from the Imperial Academy in St Petersburg was one of the largest Short had received. The Russians showed that they had kept abreast of scientific advances – eighteen of the twenty-one ordered telescopes were to have achromatic lenses. The expeditions across the Russian empire would be some of the best equipped in the world, but only if Short survived. By early summer of 1768 his strength was rapidly failing, but knowing how important the telescopes were, he forced himself to his workshop every day. Death was not far, but Short was determined to finish the instruments in time.

News did not improve throughout that summer. The Academy received a letter from the Russian astronomer who was on his way to Yakutsk, reporting that the roads towards the east had been destroyed during the previous winter. He was stuck in Siberia until conditions improved. Worried about the continual setbacks, the Academy decided to dispatch in advance the scientists employed to make natural history collections, with the astronomers to follow. Then, in mid-October, with the onset of winter, Short's instruments arrived – he had kept his promise and worked until his death to complete the order. The Academy was relieved, but still expected more instruments from France. In any case, it was now too late in the season for the astronomers to embark on their journeys. Instead the Academy decided to distribute the British telescopes and quadrants to the different teams for them to 'practise their usage' during the winter months.

By the end of January 1769, the seven astronomers who had waited in St Petersburg were finally ready to leave. Three teams were heading south: to Orenburg on the Ural river, to Guryev at the mouth of the Ural river at the Caspian Sea, and to Orsk in the south-east of Russia. Three northern locations had been chosen in Russian Lapland: Rumovsky and his team were to view the transit at Kola, while the Swiss astronomers were sent to Ponoy and Umba, also on the Kola peninsula.

On 27 January, the instructions for the expedition leaders were read out at a meeting of the Academy. Everything was minutely prescribed: how to set up the observatories, what to measure, where to determine exact geographical locations. The Academy also directed that observers should position themselves comfortably, and avoid becoming 'constrained', so as not to grow tired while watching Venus's six-hour march. Most importantly, the astronomers were to send their reports back to St Petersburg eight days after the transit at the 'very latest'.

There was only one task left for the expedition leader before they could depart: Catherine the Great had requested an audience with her astronomical army.

Catherine's Winter Palace in St Petersburg

Two days later, on a cold Sunday morning the seven astronomers set out for the Winter Palace and their important meeting. It was only a short distance from the headquarters of the Academy to her magnificent palace, just across the frozen River Neva which snaked through the city. Catherine's astronomers walked through the cold, wearing fur caps and shoes to protect themselves against the Arctic temperatures. Washerwomen had hacked holes in the icy cover of the river to soak their customers' shirts and linen. Instead of boats or carriages, brightly painted sledges flew by with 'astonishing swiftness', other visitors described, and piles of snow created artificial mounts on which people of all ages were sliding down on small trays. In the streets of St Petersburg there were merchants 'in Asiatic costume', peasants dressed in sheepskins with long beards hardened by 'clotted ice', and elegantly dressed ladies wrapped in furs. St Petersburg united, as Diderot later said, 'the age of barbarism and that of civilisation' – it was a combination of fashionable London and Paris mixed in with Asian influences and peasant culture.

Grand buildings lined the banks of the Neva – with the city's fortress and the Imperial Academy of Sciences on one side and the Winter Palace and the Admiralty on the other. The mansions of wealthy Russians and foreign merchants added to the splendour of the scene. Outside the palaces and grand buildings were fires by which servants and coachmen huddled to warm their cold limbs. As the seven astronomers were led into the Winter Palace, they unpeeled their furry cocoons to reveal their finest attire beneath. They must have looked, as one traveller described, like 'gaudy butterflies, bursting suddenly from their winter incrustations'. As they waited for the Empress, the astronomers could see the spectacular interior of the Winter Palace – painted ceilings, gilded ornaments, statues and hundreds of paintings which Catherine had bought by the cart-load in Britain, France, Netherlands, Germany and Italy. Everything exuded the finest European taste.

A Russian gentleman dressed in furs

As on every Sunday, Catherine had worshipped at noon in her palace church and was now receiving ambassadors and 'foreign gentlemen' in her drawing room for a ceremonial kiss on her hand. When the seven astronomers were presented, they saw a thirty-nine-year-old woman whom many described as still beautiful. Other contemporaries remarked upon her blue eyes which were 'expressive of scrutiny' and her charm. Most of all, she impressed with her intellect and her 'brilliant' conversation because she was familiar with such a huge range of subjects. There are no records of what exactly Catherine said to her astronomers as they bowed and kissed her hand, but they had passed their final task. They were now ready to meet Venus.

10

The Most Daring Voyage of All

After the Spanish had refused to allow the British to enter their territory in the American West, the Royal Society called a crisis meeting. What was needed to ensure the success of the British observations, it was decided, was a 'Transit Committee', dedicated to coordinating the expeditions across the globe. The man to head the committee was the indefatigable Nevil Maskelyne.

Though his observation of the first transit in St Helena had been a failure, he had succeeded in promoting himself over the past five years, and had steadily improved his position within the scientific world and the Royal Society. Never a man to let an opportunity pass, Maskelyne's first letter to the Society after his return from St Helena, for example, had underlined the 'certainty and excellency' of his lunar method of calculating longitude. Rallying further support, Maskelyne also wrote to the directors of the East India Company in a bid to convince them that his method would be of 'great benefit' to navigation. So fiercely was Maskelyne pushing his ideas that clockmaker John Harrison became increasingly worried that the astronomer might snatch away the Longitude Prize. Harrison had spent the previous three decades inventing clocks that would work reliably on a ship – in his eyes the only plausible solution to the longitude problem. Early in 1764, both Maskelyne and Harrison had been ordered by the Board of Longitude to sail to Barbados to test Harrison's timekeeper and Maskelyne's lunar tables. Harrison's fears were confirmed when he discovered that his competitor Maskelyne had been asked to judge the precision of his clock. Maskelyne pronounced Harrison's clock as not accurate enough despite its impressive results. It was the beginning of a bitter battle between the astronomer and the clockmaker which would last many years.*

In early 1765, at the age of thirty-two, Maskelyne was made Astronomer Royal, the most influential astronomical position in the country. The appointment automatically made him a member of the Board of Longitude – by this strange twist Maskelyne himself was now in charge of awarding the Longitude Prize. With his responsibilities as Astronomer Royal at the observatory in Greenwich, he would be unable to lead another transit expedition to a far-flung destination, but he would be the engine behind the global venture. Though seven other fellows were part of the Royal Society's Transit Committee, it was Maskelyne who would hold the reins over the coming years.

An ambitious astronomer who liked to be in control and oversee every detail, he no doubt felt to be the person best suited for the task. Already as a young and unknown amateur astronomer Maskelyne had written to Charles Mason (who was the assistant at the Royal Observatory at that time), instructing him on how to observe the first transit. His letter was peppered with phrases such as 'you must', 'you will' and 'I wish you would'. Now Maskelyne would do what Delisle had done for the 1761 transit – only better. He would encourage other nations to participate, propose expeditions, oversee their equipment and choose the instruments as well as selecting candidates and writing their instructions. Like a spider in a web, he placed himself at the centre of the international quest, keeping everything under his control.

Five days after the initial crisis meeting, the newly formed Committee discussed in great detail where to send the British expeditions. Two days later, on 19 November 1767, Maskelyne reported their suggestions to the other fellows of the Royal Society. The British should dispatch someone to the Prince of Wales Fort in Hudson Bay (today's Churchill in Canada), he said, as well as to Vardø and the North Cape at the Arctic Circle – 'except we learn that the Swedes or Danes will

---

* Maskelyne continued to criticise Harrison's clocks, favouring his own lunar method. The Longitude Prize was not given to one individual, but over several decades there were many recipients with Harrison receiving the greatest share. Maskelyne was not awarded, but Tobias Mayer was granted £3,000 posthumously for his lunar tables.

undertake to make this Observation'. Having covered the north, the Committee turned their attention to the most important expedition: Britain had to be in charge of the South Sea observations – the boldest and most audacious of all the voyages. They would need an expedition vessel, experienced and brave astronomers and a man who could lead his crew into the vastness of the mostly uncharted South Pacific Ocean.

This would require greater funding than the first transit had, and so the Committee proposed 'that the government be petitiond' to equip the expedition. In order to arrive there in time, they insisted, the astronomers would have to round Cape Horn by 'about Xmas 1768'. Maskelyne and his committee colleagues had been thorough and were not going to waste any time. At the same meeting they also provided lists of the necessary instruments and suggested potential candidates, including Charles Mason and Jeremiah Dixon, who were still in the North American colonies, where for the past four years they had been surveying the eponymous Mason–Dixon Line.

The next weeks rushed by at a hectic pace. In the day Maskelyne worked on the preparations for the transit expeditions and during the cold nights he observed the skies from Greenwich, no doubt sporting his brand-new 'observing suit' – a quilted outfit including a waistcoat, as well as trousers with all-in-one feet and an enormous padded bottom made of thick flannel and fine gold-, red- and cream-striped silk which reputedly Maskelyne's brother-in-law Robert Clive had sent from India. Described by a fellow scientist as a 'small invisible man', the chubby Maskelyne must have lost all remnants of authority when he waddled towards his observation room squeezed into the golden ensemble – the armour of an astronomer.

What the members of the Transit Committee didn't know was that the French were also planning a South Sea voyage. On 14 November, in between the Royal Society's decision to form a Transit Committee and their first meeting, Chappe d'Auteroche joined the French endeavour when he presented his outline for a South Sea expedition at a meeting of the Académie des Sciences in Paris. Following Pingré's calculations, Chappe suggested several locations and, of course, proposed himself as the ideal candidate for the voyage. Since this was a public assembly, Chappe emphasised with typical exaggeration the importance of the transit and declared with some embellishment and artistic licence that such a celestial rendezvous would not occur again for 'several centuries'. Just as with the first transit, Chappe was trying to take charge of the most important voyage. If he were to observe Venus from an island in the South Sea, it would make him the most famous astronomer in the world. But with Spain claiming possession of some of the islands there, the French thought it wise to obtain permission from their old ally first.*

Ignorant of the French plans, the Council of the Royal Society assembled three weeks later in London, on 3 December 1767, for a progress report from the Transit Committee. It was one of the most momentous meetings in the Society's history. After long and detailed discussions which lasted several hours, they voted to mount two expeditions, to Hudson Bay and to the South Sea. Maskelyne presented a list of potential observers and offered to write to them. He was also instructed to contact the Swedish astronomer Pehr Wilhelm Wargentin 'to know at what places they intend to observe', as well as sending a list of instruments to Stockholm, detailing the kind and size of telescopes the British were intending to use – to ensure that the results of the transit observations would be more easily comparable.

Over the next weeks the fellows of the Royal Society interviewed and appointed candidates for their expeditions. Funds were discussed and salaries negotiated. Once again they asked the East India Company to send out instructions to their employees on how to observe the transit. They received a letter from the Imperial Academy in St Petersburg, detailing the Russian expeditions and Catherine's interest in the subject.

In January 1768, as Catherine's astronomers met the Empress in the Winter Palace and as Mason and Dixon handed over the completed map of their boundary line in America, excitement about the transit was mounting in London. Newspapers reported the event and instrument-makers began to advertise improved telescopes with which to view it. Maskelyne worked on his instructions, which laid out the intricacies of the transit observations so that even amateurs could follow them. He

---

* After the Spanish refused the British request to send an observer to California, the British decided not even to ask for any permission. They would just sail into the South Sea, no matter what the Spanish said.

included explanations of the importance of the entry and exit times, different types of telescopes, how to regulate a clock and fix the instruments to the ground, as well as how to smoke different pieces of glass in shades from light to dark so as to be prepared for any weather conditions or 'flying clouds'. The instructions brought together his theoretical knowledge as an astronomer with the practical expertise gained from his own observations in St Helena.

The voyage to the South Sea required much planning and petitioning. The Royal Society needed money for instruments, lodging, food and salaries but they also required a vessel. As with the first transit, they decided to apply for money from the Crown, only this time asking for much more. To make matters worse, the Royal Society's own funds had been raided. One of their clerks had embezzled £1,500. While they had 'been attentive to what is to pass in the Heavens', one fellow informed a friend in France, the clerk had 'run away with our Money upon Earth'.

For weeks the fellows worked to draft their petition to the king. It was finished in mid-February 1768, just as Le Gentil was leaving Manila on a ship bound for India. The document stressed the need to uphold the honour of the nation and cited the potential benefits to navigation. There was no 'Nation upon Earth' superior to Britain, they claimed, and the country was 'justly celebrated in the learned world'. If the British failed to view the transit in the southern *and* northern hemispheres, the Royal Society wrote to King George III, 'it would cast dishonour upon them'. This honour, however, would cost the Crown a staggering £4,000, an amount which did not even include the costs of the necessary vessels and crews.

While the Royal Society anxiously awaited the king's reply, an impatient Maskelyne continued with the arrangements as if they had already been granted funding. With only sixteen months until the transit, he felt that there was no time to waste. The Hudson's Bay Company had warned that there was little timber in the region around the Prince of Wales Fort and had advised the Royal Society to prefabricate an observatory in England. Luckily one of the fellows was an engineer and provided drawings for a tent-like octagonal observatory made of timber and canvas. Maskelyne discussed measurements, materials and estimated costs with the carpenter in Greenwich, to ensure that the little building could easily be stored on the ship and erected at the site.

The portable observatories that British astronomers used would have looked similar to this one from Diderot's *Encyclopédie*

Time was running out because of the limited shipping season to Hudson Bay. The selected astronomer would have to leave by the end of May 1768 in order to reach his destination before ice cut off the Prince of Wales Fort until the next summer – the observer would have to overwinter there in order to view the transit in June 1769. It was of course Maskelyne who recommended a candidate for this expedition: William Wales was a young astronomer whom he had employed for the past three years to work on the *Nautical Almanac*, his latest bid to promote the lunar method. The *Nautical Almanac* provided sailors with times for eclipses of Jupiter's satellites and positions of the sun, but most importantly with the predicted lunar distances for the Greenwich meridian for each month – providing easily accessible tables from which navigators could work out how many degrees they had moved from Greenwich.* Wales would travel on a Hudson's Bay Company supply vessel

---

* Maskelyne produced forty-nine issues of the *Nautical Almanac* which became popular with navigators because it abolished the need for complicated calculations. It eventually established Greenwich as the prime meridian.

as the Prince of Wales Fort was retained by the Company to protect their interests in the fur trade with the Native Americans (and against the French). The directors had assured the Royal Society that they would provide passage, accommodation, food and any other assistance to the observers for a one-off payment of £250.

Everything was progressing smoothly. As Maskelyne was calculating the exact tonnage and measurements of the portable observatory (the Hudson's Bay Company had requested this information in order to find space on their ship), the secretary of the Admiralty ordered the Navy Board to prepare a vessel for the South Sea expedition. 'His Majesty has been graciously pleased', the secretary wrote, 'to defray the expenses' for the purchase of a vessel. With his interest in science, King George III had been quick to reach a decision and, less than three weeks after reading the Royal Society's petition, had ordered the search for a suitable vessel. Ships were inspected and rejected but none seemed quite right. One had to be altered too much, another couldn't 'stow the quantity of provisions required', others were at sea and wouldn't return in time.

On 24 March 1768, the king informed the Royal Society that he had granted not only the provision of a vessel for the South Sea expediton, but also £4,000 from his personal funds. The fellows were jubilant. Never before had so much money been spent on a scientific project. Less than a week later, the Navy purchased a so-called 'cat', a ship that had been built to carry coal from north-east England to London, and named it *Endeavour*. Preparations for the South Sea voyage could begin in earnest.

The *Endeavour* was refitted in Deptford – re-rigged and covered in a mixture of pitch, tar and sulphur to protect the timbers against tropical water worms. Overseeing the alterations was James Cook, who had been chosen by the Navy to head the expedition. Born in 1728 in Yorkshire, Cook had begun his career at sea as a deckhand on a similar 'cat', transporting coal along the English coast. He had later joined the Navy, fought in the Seven Years' War and charted Newfoundland. A quiet and private man, Cook was known to be cautious, but also willing to take calculated risks when necessary. He was involved in every detail of the preparations and would not only act as the captain of the *Endeavour* but also as an observer. Cook was a skilled cartographer and astronomer, a well-regarded marine surveyor, able mathematician and experienced sailor – but he would return from the *Endeavour* voyage as a respected leader, celebrated discoverer and a hero.

Accompanying him, as the astronomer for the South Sea expedition, would be Charles Green, Maskelyne's former assistant. Green was a thirty-three-year-old Yorkshireman who had worked for several years at the Greenwich observatory, where he had observed the first transit.* Like many of the other assistants there, he was keen to leave his tedious job, and had a liking for sea voyages. In 1763, he had accompanied Maskelyne to Barbados where they tested John Harrison's clock for longitudinal determinations. Shortly afterwards, Green left the Royal Observatory to join the Navy. When he heard that the Royal Society was mounting several expeditions to view the second transit, Green had gone to one of the meetings to propose himself as an observer and to introduce his terms and conditions – for £300 per year, he had declared, he would be delighted to 'go to the Southward'. In fact, he was so keen to leave that he finally accepted a yearly salary of £100. Much less than he had first asked for, it was nonetheless more than Cook's one-off payment of £100 for his astronomical observations. Cook might have been the commander of the expedition but he was only the second observer. For the Royal Society the lead astronomer was the most important person on the vessel.

Maskelyne put together his instructions for Cook and Green, and oversaw the construction of the portable observatory. Upon their arrival, Maskelyne insisted they should set up their equipment, 'without any loss of time'. It was important that every observer practised with each instrument for as long as possible in order to take the transit times with the greatest precision.

At the same time Maskelyne was finalising the last details of William Wales's journey to Hudson Bay, advising the young astronomer how to store the observatory and instruments on board. Almost exactly a year before the transit, on 29 May 1768, Wales left London after a midnight visit to the Greenwich

---

* Charles Green was also William Wales's brother-in-law – the astronomer who was appointed to observe the transit in Hudson Bay.

observatory where Maskelyne gave him his written instructions and a last-minute briefing.*

Everybody was frantically busy. While Maskelyne simultaneously worked on the Hudson Bay and South Sea expeditions, Cook concentrated on his ship. He requested eight tons of ballast because she 'Swims too much by the Head' and discussed the employment of a crew with the Navy, as well as the beer and spirit allowances. Meanwhile the Royal Society was debating where exactly in the South Sea their astronomers should observe the transit. Luck was on their side. Just at that time a British captain named Samuel Wallis returned from a long voyage with news of an island in the South Sea populated by peaceful natives and stocked with abundant food. Tahiti, or King George Island as it was then known, seemed the perfect location.

As the astronomers prepared the transit expeditions, riots broke out in England – affecting many trades essential to provisioning a sea voyage. Food shortages and wage cuts, combined with the imprisonment of radical journalist John Wilkes, had incited Londoners to take to the street. Coal heavers, tailors and even prostitutes in brothels (for the 'cuts' that pimps and tavern-keepers were demanding) went on strike, but also sailors, coopers, weavers and glass-grinders.

In the midst of the chaos, Cook was pushing for more provisions and equipment. Every day, several letters went back and forth – Cook requested green cloth for the floor of the Great Cabin, more 'Surgeons necessarys', guns, 'a Machine for sweetening foul water' and 'portable soup'. Diligent and determined to be prepared for all eventualities, he was not missing anything: they needed more salt, he discussed methods of avoiding scurvy† with the Sick and Hurt Board, and ordered instruments for surveying and making maps.

---

* After the midnight meeting with Maskelyne, Wales dashed home to bid goodbye to his heavily pregnant wife Mary and their baby girl. He thought it best to send his wife and daughter from Greenwich to Yorkshire where they could stay with his relatives during his long absence. Unsurprisingly – given the state of the roads and the badly sprung coaches – his wife gave birth somewhere along the Great North Road in Lincolnshire, halfway to Yorkshire.

† In the eighteenth century more sailors died from scurvy than at the hands of the enemy.

Only the choice of his crew – undoubtedly one of the most decisive factors for the success of such a voyage – was taken out of Cook's hands. The Navy selected the men, but when they sent a lame and infirm sailor as cook, the captain complained. Three days later a new cook was ordered to the *Endeavour* – though he seemed even worse than the first because, Cook protested, 'this man hath had the misfortune to loose his right hand'. Despite his protests the Navy insisted that the one-handed cook should remain on the *Endeavour*, no matter if the captain thought he would be of 'little Service'.

Meanwhile Maskelyne dealt with the instruments and astronomical directions, and the president of the Royal Society wrote long and detailed instructions on conduct and tasks. The crew, the president wrote, had to exercise 'utmost patience' with the natives they would encounter, since 'no European Nation has a right to occupy any part of their country'. The primary object of the voyage was the observation of the transit of Venus, he reminded Cook.

Like Catherine the Great, the Royal Society also broadened the objectives of the expedition in the spirit of the Enlightenment. They issued instructions on making ethnographical observations and botanical, mineral and animal collections. Cook and his crew were to collect specimens and acquire knowledge in order to make sense of this new world. Once Venus was observed, the *Endeavour* was not to return immediately but Cook should also make it his task to discover 'a Continent in the Lower temperate Latitudes' – Terra Australis Incognita – 'the unknown land of the South'.* The transit expedition had become a broader scientific exploration with colonial and economic undertones.

Finally, on 25 August 1768, Cook and his crew were ready to leave. The wind billowed the sails as the *Endeavour* veered south and ninety-four men headed into the unknown to see Venus traverse the face of the sun. It was a clear afternoon and the vessel was filled to the brim. Over the past few days the sailors had carried so many provisions on board that there

* At that time people believed in the symmetry of the world, thinking that there was a huge southern landmass to balance that in the northern hemisphere.

was hardly any space to move. In addition to eight tons of ballast, there were twenty tons of biscuits and flour, 10,000 portions of salted beef and pork, as well as 2,500 pounds of raisins amongst the supplies. Cook had loaded 1,600 gallons of spirits and 1,200 gallons of beer to keep his crew happy, and almost 8,000 pounds of sauerkraut to test its effect against scurvy. There was timber, ropes, nails, tools and canvas for emergency repairs, as well as a vast array of mirrors, beads and other trinkets as presents for the natives.

The collection of astronomical instruments was over-whelming – ranging from quadrants and astronomical clocks to several of the best telescopes available – and of course there was the portable observatory that Maskelyne had organised. Green had already been on board for almost three weeks, ensuring that the instruments were safely packed and stored. Cook had watched in dismay when the botanist and wealthy landowner Joseph Banks arrived with an entourage of eight, comprising servants, artists and another botanist, as well as an alarming array of luggage which included hundreds of specimen bottles, papers, plant presses, microscopes and a library of more than 100 natural history books in addition to drawing tables and a desk for Banks. Everything was packed in twenty large wooden cases. 'It almost frighten[s] me', Banks himself admitted. He had spent a staggering £10,000 on his and his team's passage on the *Endeavour* and prided himself on being the 'first man of Scientifick education who undertook a voyage of discovery'. Once Cook had seen the paraphernalia, he had ordered the carpenters to make yet more changes to the cabins in order to fit everything in.

Cook had done his utmost to ensure the success of the voyage. They were all risking their lives for the transit. Many were leaving behind families – astronomer Charles Green was married and three weeks previously Cook had kissed his chil-dren and his heavily pregnant wife Elizabeth goodbye.* At least they were all in 'excellent health and spirits', Banks said, and 'perfectly prepard (in Mind at least)' for their long and dangerous voyage.

---

* Ten days after the *Endeavour* sailed, Elizabeth Cook would give birth to a son whom Cook would never see because the baby died within a month.

11

Scandinavia or the Land of the Midnight Sun

'I do believe, this time there will be fewer decisive observations', the secretary of the Royal Academy of Sciences in Stockholm, Pehr Wilhelm Wargentin, wrote to his colleague Anders Planman. Once again the untiring Wargentin was taking charge of the Swedish contributions, but unlike the British, French and Russians, he was less optimistic about their success. He worried that the most important destinations for this transit were even further away and harder to reach than they had been in 1761. This made the Swedish observations in the far north of the kingdom even more vital because it was 'the only place in Europe' from where to see both the beginning and the end of the transit. The main problem for the Swedes, however, was that the Academy in Stockholm was almost broke* and Wargentin was at a loss about how to underwrite the expeditions. With plenty of skilled astronomers in Sweden, it would be of 'everlasting harm and shame' if they had to ask the English or French to undertake observations in Scandinavia, he lamented.

As in 1761, Wargentin was up to date with the transit preparations elsewhere. Though far away from the European scientific centres of London and Paris, he was better informed about the latest developments than most others. Wargentin maintained an international correspondence that seemed superhuman, writing during his time as the secretary of the Academy several thousand letters. With steely patience and exceptional organisational skills, he had quietly turned around the fortunes

---

* The main income of the Academy was an almanac which due to increased paper and printing costs was not as profitable as it had been a few years previously.

of the Swedish Academy. Under his guidance it had become a recognised member of the European network of learned societies – all part of the 'desire to increase our prestige and our glory', he explained.

Wargentin acted as the chief point of contact for the international community and presented the letters which he received at the meetings of the Academy. The English and the French, he told his colleagues, reported on their own transit expeditions but also stressed that the success of these observations relied on corresponding data from the north, ideally taken in Torneå in Lapland. Wargentin in turn urged his colleagues in Sweden to cooperate, otherwise none of these results would be useful. The discussion went back and forth – with little money in the coffers, some Swedish scientists suggested immediately writing to Paris and London to inform their colleagues that they would have to dispatch their own observers. Wargentin was outraged. He hadn't spent the past two decades dragging Sweden into the international exchange of learning to fail during the most important scientific event of the eighteenth century. The only solution was to appeal to the king. It was, Wargentin said, a matter of national honour.

In early 1767 Wargentin composed a long petition to King Adolph Frederick, explaining the importance of the endeavour for all nations that were engaged in 'science, trade and seafaring', as well as drawing attention to the French and English efforts. As the transit had been visible everywhere in the country in 1761, it had been relatively cheap to organise the observations in Sweden – many astronomers had simply watched it from their homes or nearby observatories – but now the situation was very different.

'There was no other place in the world', Wargentin said, 'that would be as suitable' as Lapland – a site that would act as the perfect counterpart to the British South Sea observations. In case of clouds or illness, they would need to mount at least two or three expeditions. And to underline the importance of the enterprise not only for the world of science in general but specifically for Sweden, Wargentin explained that Planman's longitudinal observations during the last transit had helped to map some parts of Lapland – an important advance in their imperial ambitions to exploit the potential riches of the northern lands. Wargentin's strategy worked. Only two weeks later, on 29 January 1767, the king granted them funds. More than two years before the transit the Swedish contributions seemed to be secured.

Efficient as always, Wargentin proposed the first expedition on the same day: he wanted to send the astronomer Fredrik Mallet* to Pello in Lapland, north of the Gulf of Bothnia and within the Arctic Circle. Wargentin knew the thirty-nine-year-old Mallet well – he was an excellent astronomer but also deeply melancholic. As the son of a wealthy factory owner, Mallet had enjoyed a comfortable upbringing but had quickly used up his inheritance and had subsequently struggled to make a living. Mallet was passionate about astronomy and mathematics but unable to find a paid job. As a young man he had travelled through Europe, meeting famous thinkers and scientists. These connections had fed his thirst for knowledge but also his taste for the excitement of metropolitan life in London and Paris. After his return from this scientific tour, Mallet had become even more miserable than before, feeling confined by the provinciality of Uppsala where he volunteered at the observatory. With clockwork regularity, he had been ignored again and again when paid positions had become available – so often that Wargentin admitted that even a man 'plagued less by melancholy than Mallet' would have despaired.

In 1761, the troubled astronomer had watched the first transit at the Uppsala observatory, declaring with his penchant for melodrama that he would give up astronomy and have himself 'hanged' if he failed to see Venus. Luckily for him the weather had been obliging and Mallet had succeeded in viewing the entire transit. Worried as always, he had also taken the precaution of monitoring the street in front of the observatory so that 'no horse would pass' to avoid any disturbance of the delicate instruments during the event.

Since then Mallet had grown increasingly gloomy and impatient. He regularly threatened to abandon astronomy, and when he was yet again overlooked for a position, wrote to Wargentin,

*Fredrik Mallet was not related to the Swiss astronomer Jacques André Mallet who was dispatched to the Kola peninsula by Catherine the Great.

'I am incapable of a good mood'. He wanted to leave Uppsala. With these regular complaints in mind, Wargentin immediately thought of Mallet while considering possible candidates for the northern expeditions. Though Lapland was not London or Paris, the journey would distract the unhappy astronomer, get him out of Uppsala, give him a sense of purpose – and a salary.

By spring 1767, as Catherine the Great was ordering the Russian expeditions, the Swedish efforts were taking shape too. Mallet agreed to travel to Pello, and Planman, who now worked as a professor at the University of Åbo, was ordered by Wargentin to observe the transit once again in Kajana. Planman, who had obsessively calculated and recalculated the solar parallax over the past years, welcomed the assignment. He did not want to miss the opportunity to see Venus a second time. Within weeks letters were zigzagging across Europe, spreading the news of the Swedish plans. Wargentin's report was read to the Imperial Academy in St Petersburg, which Stepan Rumovsky – in an attempt to underline how much Russia was involved in the global enterprise – copied and dispatched to the Royal Society in London.

Maskelyne was relieved and the Russians were convinced that their own expeditions, combined with those to Pello and Kajana, would ensure the success of the northern observations. 'One or other of these Stations will produce a compleat observation', Rumovsky wrote to the Royal Society. The British expressed some doubts about Wargentin's choice of locations, suggesting there was a danger that the sun would be too low above the horizon in Pello and Kajana. But they were satisfied that Mallet and Planman were diligent observers, exerting 'their utmost to give us the best observations they are able to make'.

In February 1768, a few months before Mallet's departure to Pello, the Academy in Stockholm finalised the details of the expeditions and divided the funds which the king had granted between Mallet, Planman and an amateur astronomer who lived in Torneå at the northern end of the Gulf of Bothnia. Mallet, who was venturing the furthest, received more than half.

In response to the many disappointments of the previous observations, every country involved with the second transit expanded the tasks of the travellers by requesting that they go in the name of science, and not just astronomy. Fertile lands, valuable crops, opportunities for settlement and conquest – in short an expansion of empire and the exploitation of soils, plants and minerals – added another dimension to the astronomical expeditions. With important implications for trade, Mallet was now asked by the Swedish Admiralty to survey the coastline of the Gulf of Bothnia in order to update the existing naval maps and 'to determine the principal places and harbours'.

During the summer of 1768, Mallet studied the few existing maps of the Gulf of Bothnia, while the Academy in Stockholm appointed three members to prepare and adjust the instruments that Wargentin had bought from England for the first transit. Meanwhile Planman was still obsessing about the parallax calculations, writing to Wargentin that he had come up with yet another formula and that he was 'indescribably satisfied' with his new method.

In August, Mallet left Uppsala on his long journey to Pello, but only a few days later his assistant became so ill that they were forced to wait for several weeks in Öregrund only fifty miles from Uppsala at the south-western end of the Gulf of Bothnia, while he recovered. It was the most 'deplorable' place according to Mallet. He rented a small open boat to survey the coastline, but once again, nothing went to plan. Caught up in storms, he spent several nights lying on the bottom of the boat, exposed to wind and rain. Without his 'courage and perseverance', Mallet said, 'I should have given up'. He quickly worked himself into a bad mood and sent a downhearted letter to Wargentin, expressing doubts of reaching his destination as winter was fast approaching. He still had most of the journey before him. Blizzards, snow and storms would make travelling slow and uncomfortable. Stuck in Öregrund, Pello suddenly seemed unreachable to him.

While Mallet and Planman were appointed as the main astronomers for the Swedish observations in the far north, another major Scandinavian expedition was organised – this time under the aegis of the Danish king. Christian VII had instructed his ambassador in Vienna to ask Jesuit priest and astronomer Maximilian Hell to observe the transit of Venus at the Crown's expense in Vardø, a small island in the Barents Sea. Finally, someone had listened to Maskelyne, who for so long had recommended sending observers to the most north-eastern point of Norway. Wargentin had ignored Maskelyne's pleas, but nineteen-year-old Christian had taken up the challenge. Vardø was a Danish garrison (Norway was under Danish control) and the king had ordered his commander there to provide lodging and assistance to the astronomers.

Like Catherine the Great, Christian VII was trying to portray himself as an enlightened ruler through patronage of the sciences. So important had the advancement of science become that Christian requested to be made a fellow of the Royal Society in London. He would be 'very much flattered to be chosen', he informed the fellows. Taking full advantage of the expedition to Vardø, he also added a botanist to the payroll in order to make a collection of northern plants.*

Science came before all. Christian VII placed more importance on intellectual excellence and scientific expertise than on religion and the law. At a time when Jesuits were banned from even entering Protestant Denmark, the invite to Hell to conduct the country's contribution to the most important astronomical observation of the century, clearly displayed the king's scientific credentials.

Hell was happy to oblige. As the director of the Royal Observatory in Vienna, he had viewed the first transit there – alone in a tower as the observatory had been packed with too many visitors – but the 1769 transit could not be observed from Austria. There was no better viewing location in the northern hemisphere than Vardø, he believed. In the Land of the Midnight Sun the entire transit would be visible. And as the region in the high north of Norway was completely unknown to the world of science, the expedition would also give Hell the opportunity to investigate its climate, soil and the indigenous population.

Unlike the more gregarious Nevil Maskelyne and Chappe d'Auteroche, forty-eight-year-old Hell was not hungry for fame or adventure. He was a humble man whose passion for astronomy was matched only by his love of God. For him astronomy revealed the wonder of God's creation. It was his belief that the discovery of the distance between the earth and

---

* The botanist was Jens Finne Borchgrevink who had recently finished his studies with the most important Swedish botanist of his time, Carl Linnaeus.

the sun would only add to humankind's knowledge of the divine architect's glory.

On 28 April 1768 Hell began his long and arduous journey from Vienna towards the frozen world of the Arctic Circle. Like the other transit astronomers, he did not travel lightly. His scientific instruments alone included two quadrants, two large pendulum clocks, several telescopes (of which the largest was ten and a half feet long), a small library of scientific books, reams of paper and many bottles of ink, as well as olive oil, chocolate, coffee and tea. Hell planned to take the land route from Vienna to Trondheim on the west coast of Norway from where he had been advised to catch a boat to sail around the northern tip of the country. The expedition would have to arrive before winter clasped its icy grip around the coast, but first Hell had an appointment with the king of Denmark in his castle near Lübeck.

Together with his assistant János Sajnovics, a servant and a dog called Apropos, Hell travelled from Vienna through Prague, Dresden, Leipzig and Hamburg to Lübeck. Like Delisle's journey to Russia four decades previously, Hell and Sajnovics turned this part of their expedition into a Grand Tour of scientific minds. Everywhere they met astronomers who had watched the first transit, exchanging information and tips for the next one. Sajnovics, whose diary reveals a man with a keen eye for his surroundings, enjoyed the earthly pleasures of life too. Where the wiry Hell was frugal and ascetic (who ate little and fasted every Saturday), Sajnovics savoured good food and comfortable beds – which sometimes did not make them the best of travel companions. In Dresden, Sajnovics complained that people preferred to spend their money on clothes and gardens rather than food and drink, in Hamburg he criticised the houses for having too many windows, and he said that the roads to Lübeck were the worst he had ever encountered. Hell insisted on travelling simply and the carriage he organised, Sajnovics sighed, was a 'wretched' cart with no springs. This rough travelling wrecked the instruments, and when they opened their luggage they found the barometers and thermometers broken – with tiny pearls of mercury floating in their clothes – but the telescopes and quadrant emerged undamaged.

To Sajnovics's dismay they had to stay in a 'miserable' tavern when they arrived at the king's castle just outside Lübeck. With the heavenly clock ticking, they impatiently waited several days for an audience with their patron. When they finally met, on 1 June 1768, Christian VII welcomed Hell warmly. 'I'm pleased', the Danish king told the Jesuit, that 'such a famous astronomer agreed to make this important observation'. The king assured the astronomers that they would receive all the help they needed for their long journey. They were to proceed as quickly as possible to Trondheim (where they would meet the botanist). There was not much time. It had taken them a month to cover the 600 miles from Vienna to Lübeck. Now they had only two months to travel with all their instruments to Trondheim – a journey of 1,000 miles, much of which was across the rough and mountainous emptiness of Norway.

By 3 June 1768, with exactly one year to the transit, an impressive six expeditions to the Arctic Circle had been organised by Sweden, Denmark, Russia and Britain. The king of Denmark was paying for Hell's voyage to Vardø, from Stockholm Wargentin was managing Mallet's expedition to Pello in Lapland (paid for by the Swedish king), while Catherine the Great had employed astronomers to travel to three different locations on the Kola peninsula in Russian Lapland. In London Maskelyne was planning yet another voyage – to the North Cape at the northernmost tip of Norway which he hoped King George III would fund.*

All these northern observations were essential to the success of the transit enterprise. They were to form counterparts to the *Endeavour*'s dangerous voyage to the South Sea.

---

* On 2 June 1768, Maskelyne had also welcomed King George III and Queen Charlotte to the Royal Observatory. Having given the Royal Society £4,000 for the transit expedition to the South Sea as well as having ordered a new observatory for himself at the old hunting park in Richmond and Kew, George III was much interested in the transit and most certainly requested an update from his Astronomer Royal.

The North American Continent

The North American continent became an important location for astronomers during the second transit. The thirteen colonies along the eastern seaboard would be able to observe the beginning of Venus's march across the sun from the early afternoon until sunset, missing only her exit. But in Hudson Bay in the far north the entire transit would be visible, as it would be in the west in the Spanish territories of Mexico and California.

When the British dispatched William Wales at the end of May 1768 to spend a long winter in Hudson Bay, they were the first to organise an expedition to the North American continent. The French, who also planned to send a team, were lagging behind. Chappe d'Auteroche, at the Académie in Paris, had petitioned to mount an expedition to the South Sea, but the Spanish had refused them permission. Instead Carlos III had offered the French passage on a Spanish ship to Mexico in order to observe the transit in California. No French vessel was allowed to enter Spanish territory unescorted – only if the Spanish could control the voyage would France be able to participate. Luckily there were Spanish merchant fleets travelling to Mexico, so the gesture would cost not much. But under no circumstances was Carlos willing to pay for an expensive and dangerous voyage under the Spanish flag.

Chappe's dreams of fame in the South Sea were dashed, but the French quickly turned their attention to the American West instead. They would at least be able to view the transit from California (where the British were not allowed to travel). Over several months letters raced back and forth between the Académie and the Spanish. Suggestions for observers were made, then overruled. It can't have helped that Delisle had seemingly turned away from the scientific world which for so many decades had been the centre of his life. After the first transit, the old widower had first found solace in religion and then with an Ottoman princess – at least according to the rumour mill in Paris. One astronomer said Delisle had become so 'attracted' to the daughter of the late Sultan of Constantinople, Ahmed III, that he 'could not part with her'. With Delisle preoccupied with other matters and Lalande more concerned with theory, predictions and parallax calculations than with the practical aspects of mounting expeditions, there was no mastermind behind the French effort, and the Académie struggled with the organisation.

While the French preparations were dragging on, the Spanish finalised their part of the deal. Two naval officers with astronomical skills were assigned to the venture. They were to meet a French team in Cadiz and sail to Veracruz on the west coast of the Gulf of Mexico from where they would have to cross the whole of Mexico to reach Baja California on the Pacific Ocean. Carlos III ordered them to calculate their exact geographical positions along the way and to keep a detailed diary. The officers were to observe the transit, but also had to watch the French and 'never to separate themselves' from them, the Spanish king insisted.

While the European powers divided the North American continent according to their colonial reach and political alliances, the American colonists on the eastern seaboard were not going to let their mother country take all the credit. As William Wales sailed towards Hudson Bay at the end of June 1768, the Americans were planning their own observations. On 21 June, thirteen scientifically minded Philadelphians met at the State House to discuss the transit. They pored over the projections of Venus's path and the calculations that predicted where and when the planet would appear on the face of the sun.

The thirteen men were members of the American Philosophical Society (or APS) which had been founded by Benjamin Franklin and some Philadelphian friends in 1743. The early founders had hoped to emulate the British Royal Society when they established their own scientific forum for 'Promoting Useful Knowledge', but Franklin had quickly discovered that the members were all 'very idle Gentlemen' who failed to participate in any scientific endeavour. For three decades nothing had happened at the APS. The transit of Venus, they hoped, was going to change this.*

Scientists in the colonies were aware that their European colleagues were not impressed with the progress of science in America. France, in particular, looked down on them. Only a couple of years previously the most famous French naturalist, Georges-Louis Leclerc, comte de Buffon, had published his offending ideas on America's 'degeneracy', in which he claimed that anything that was 'transported' to America – plants, animals and humans – failed to thrive there. One French thinker even claimed that America had never produced a 'man of genius in a single art or a single science'.

With American newspapers reporting on Cook's *Endeavour* voyage and the Russian expeditions, the colonists realised just how interested the world was in this global enterprise. 'Much depends on this important Phenomenon', the members of the APS insisted. Like Catherine the Great, who utilised the transit to cast Russia as an enlightened European nation, the Americans believed that if they could organise several observations, the world would hold them in higher regard, rather than belittling them as backward-thinking farmers.

When Franklin had first proposed the APS, he had argued that the colonies were ready for scientific projects, because 'the first drudgery of settling new colonies . . . is now pretty well over'. And Philadelphia was the ideal place for such an enterprise. With 30,000 inhabitants, it was the largest city in the North American colonies, fed by a lively trade with Britain. Ships criss-crossed the Atlantic loaded with American crops, cotton and tobacco for Britain, and luxury goods as well as staples such as cloth, paper and nails for the colonies. Philadelphia was a neat city, with streets named after native trees and laid out in a regular grid. The sidewalks were paved and lights lit up the dark streets during the nights. The city had a university, the first American subscription library, and,

* There were in fact two competing societies which tried to take the baton from the early APS in the late 1760s. After an acrimonious battle, the two formally merged on 2 January 1769, under the name American Philosophical Society.

thanks to Franklin, had become the hub of the fast-growing postal service in the colonies. Wealthy Virginians with their European tastes might have felt a snobbish affection for being the oldest colony, but Philadelphians were proud of their city's burgeoning intellectual and commercial life.

Most excited about the prospect of America's contribution to the transit project was thirty-six-year-old David Rittenhouse, a member of the APS and self-taught instrument-maker and astronomer from Norriton, some twenty miles north-west of Philadelphia's city centre. Rittenhouse had been fascinated by mechanics and astronomy since childhood. As a young boy he had shown a greater interest in ornamenting plough handles, fences and barn doors on his father's farm with carvings of astronomical constellations than in agriculture. By the age of nineteen he had opened a clock-making workshop on the farm while continuing his other scientific studies.

Rittenhouse's predictions and calculations formed the basis of the American selection of viewing stations. Several members of the APS had offered their services as observers because 'the Beginning and a great Part of it will be visible at Philadelphia, if the weather should be favorable'. By the end of the APS meeting, the thirteen men had decided that one group should view the transit from Rittenhouse's farm and another from Philadelphia.* Rittenhouse was asked to make 'the necessary Preparations' because he was the most suited man for the task. There was no one in the colonies who combined so much theoretical astronomical knowledge with practical skills.

Only three months before the transit meeting in June 1768, Rittenhouse had greatly impressed his scientific friends with an orrery that showed the movements of the planets. Unlike other orreries which were designed simply to show how heavenly bodies rotated around the sun, Rittenhouse had constructed his complex mechanism with such staggering accuracy that it could simulate any astronomical constellation at any given date between 4000 BC and AD 6000, including solar eclipses as well as transits of Venus and Mercury. It was a magical piece of mechanics and an astronomical masterpiece. Rittenhouse,

---

* They later added another location in Delaware.

Thomas Jefferson later said, had created a world that 'approached nearer its Maker than any man who has lived from the creation to this day'.

Rittenhouse was not the only scientist to oversee the colonists' transit observations. Benjamin Franklin was acting as an intermediary between Britain and America. The sixty-three-year-old was living in London as the agent of the Pennsylvania Assembly, but had quickly become a sort of unofficial ambassador for the colonies during the increasingly tense state of relations after the Stamp Act crisis of 1765. Though his sojourn had a political purpose, Franklin also immersed himself in Britain's thriving network of thinkers, philosophers and scientists.

A fellow of the Royal Society and a man with an insatiable curiosity, Franklin had become deeply involved in the transit expeditions. He was fascinated by nature and investigated a vast range of subjects, from 'Snuffling' air from the bottom of ponds to measuring the temperature of the ocean during his transatlantic voyages in order to chart the course of the Gulf Stream. He loved being in London, which he regarded as the nexus of scientific enquiry. Franklin toured the coffee houses and clubs, attended the meetings of the Royal Society and was courted by Britain's greatest thinkers. 'America has sent us many good things, Gold, Silver, Sugar, Tobacco, Indigo &c.', the Scottish philosopher David Hume wrote to Franklin, 'but you are the first Philosopher, and indeed the first Great Man of Letters for whom we are beholden to her'.

During his previous visit to London, Franklin had been a member of the Council of the Royal Society as they were preparing Mason and Dixon's expedition to Bencoolen, and Maskelyne's to St Helena, for the first transit. Franklin had sent pamphlets, books and reports on the results to his friends in the colonies. When he briefly returned to Philadelphia in 1762, his scientific friends in London had provided him with information about the second transit, encouraging him to persuade the colonists to participate.

On his subsequent return to London, Franklin was once again elected to the Council of the Royal Society and had since then taken part in the preparations for the British expeditions to observe the second transit. He had listened to the suggestions of the prospective candidates and had been involved in the arrangements for the Hudson Bay observations. Franklin had attended meetings of the Society when letters from Catherine the Great and the Imperial Academy in St Petersburg were read and had worked on the drafts for the petition for funds from King George III. He had seen the designs for the portable observatories and Maskelyne's long lists of instruments. In May 1768, Franklin had met James Cook and Charles Green when they were appointed to sail to the South Sea and had debated with the other fellows the best choice of destination. With his scientific knowledge and London connections, Franklin was invaluable in assisting his fellow colonists to organise their own transit observations.

He was not only helping the APS in Philadelphia but also his acquaintance John Winthrop in Massachusetts, who had successfully observed the first transit (albeit for only a short period) in Newfoundland. After a devastating fire at Harvard University had destroyed the telescopes there, Winthrop had asked Franklin to purchase a new one in London from James Short. Franklin duly ordered the instrument. On 2 July, only ten days after the APS transit meeting in Philadelphia, Franklin wrote a letter to Winthrop explaining why he had been unable to dispatch it: Short was dead. Though Short had finished his orders for Russia, Winthrop's telescope was now stuck in the complicated probate of his estate. Franklin had put forward a claim for the instrument but had to wait until the executors of Short's will had settled his affairs.

Winthrop's order for equal altitude and transit instruments from London instrument-maker John Bird had also been delayed. There was such a 'great and hasty demand on him from France and Russia, and our society', Franklin reported, that Bird had not even started to work on the American's order. But with the instruments for the European expeditions now dispatched, Bird had promised to finish Winthrop's by the end of the following week. 'Possibly he may keep his word', Franklin wrote, but also warned, 'we are not to wonder if he does not' – with the transit less than a year away, instrument-makers in London were working around the clock.

Nevil Maskelyne, who continued to canvass for as many observations as possible, had also been discussing the colonial effort with Franklin. The Astronomer Royal was hoping that an American observer would travel to Lake Superior to view the transit from there. To encourage Winthrop, Franklin forwarded some of Maskelyne's letters as well as his instructions. Desperate to show that the North American colonies were not a wilderness hinterland, but equally capable of contributing to science as the European countries, Franklin reminded Winthrop how important and highly regarded the Newfoundland expedition in 1761 had been.* The observations of the second transit would be a 'great honor' for the colonies – but the Americans had to hurry if they wanted to participate. 'If your health and strength were sufficient for such an expedition', Franklin wrote, 'I should be glad to hear you had undertaken it'.

ICEBERGS.

While the colonists were discussing where best to view the transit, the British were about to arrive in North America. Astronomer William Wales was the first to reach his destination. The journey had been longer than anticipated because they had been delayed by bad weather. They had encountered

---

* Following his 1761 transit observation, Winthrop had been made a fellow of the Royal Society.

icebergs which rose next to the vessel like islands, with glittering spires that towered high above their main mast. At first Wales thought them to be 'most romantic', but then sailing had become perilous. Enveloped in thick fog, the captain blindly manoeuvred his ship through the treacherous labyrinth of floating ice. 'Our situation must be allowed to have been truly dangerous', Wales drily noted in his journal.

At the end of July they had reached the entrance to Hudson Bay and sailed towards Prince of Wales Fort at the mouth of the Churchill River on the west coast. They anchored on 10 August. It was a bleak environment of bare rocks with little vegetation – some dwarf willows and birch as well as gooseberry bushes – but nothing, Wales noted, 'that would bear the name of trees'. Luckily they had brought their prefabricated observatory which they set up on the stone bastions of the fort. Diligently following Maskelyne's detailed orders, Wales and his assistant concentrated on their workplace rather than their own accommodation. For two weeks they slept on the bare wooden planks of their cabin floor before the carpenter finally found time to build two beds. They were pestered by mosquitoes and millions of tiny flies which, Wales moaned, were so obnoxious that it was impossible 'either to speak, breathe or look, without having one's mouth, nose or eyes full of them'. Winter would arrive soon and Wales, in particular, would suffer because he hated the cold. Ironically, the Royal Society had dispatched the one man who had emphasised in his interview that he was 'preferring a Voyage to a warm climate'. Instead Wales would have to wait for Venus for ten long, cold months.

The French, too, were slowly progressing with their own North American expedition. In August, as Wales worked on his observatory, the Académie in Paris finally assigned an astronomer to the voyage to California: Chappe d'Auteroche.*

---

* With Chappe commissioned, the Académie also tried to get permission from the Spanish for Pingré to observe the transit near the coast of Mexico. Pingré was to be part of an expedition to test the precision of the latest French timekeepers used to calculate longitude at sea. It quickly became apparent that the Spanish would 'not allow any French ship near the coast'. Without permission to go to Mexico, it was left to Pingré to decide from where to view the transit of Venus – which he did from Haiti (then called Saint–Domingue or Santo Domingo).

Disappointed that he couldn't carry the flame of fame to the South Sea, Chappe chose what he regarded as the next-best French expedition – at least he would not be part of the throng of astronomers that was heading towards the frozen desert of the Arctic Circle or Hudson Bay.

A view of the Prince of Wales Fort in Hudson Bay

To make it to California in time, Chappe had to leave Paris as soon as possible. There was much to do: he had to assemble a team and his equipment as well as arrange where, and how, to join the Spanish astronomers. Permission to enter California had only been given on condition that the transit observation was the 'sole purpose of this journey' and that Chappe would 'go in the company' of the Spanish – something that he was probably not too pleased about. But before he could set off for California, he had one task to see through: the publication of his three-volume *Voyage en Sibérie*.

On 31 August 1768, the same day as the Hudson's Bay Company's ship that had conveyed Wales sailed back to England, the members of the Académie in Paris listened to a long and detailed report on Chappe's book. Their verdict was unanimous: the publication deserved their 'merit of approval'. The French Académie might have been pleased with Chappe's effort, lavishing praise on him, but the book would also provoke the wrath of one of the most powerful women in the world: Catherine the Great. Though Chappe had promoted his book to the Académie as an account of his transit observations in Siberia, as well as a comprehensive survey of Russia's natural history, climate, soil and customs, in Catherine's eyes it was an outrageous portrait of her empire as a country populated by impoverished alcoholics and superstitious peasants. According to her, it was a 'malicious misrepresentation' which undermined her attempts to present Russia as an enlightened European nation.

So enraged was Catherine that she published her own book in response: *Antidote*, a hilariously funny point-by-point rebuttal written in French and English (and clearly aimed at a European audience). Directly addressing the author of *Voyage en Sibérie*, Catherine ridiculed Chappe over more than 200 pages: his comments about peasant women, she countered with, 'I observe you take great notice of the women . . . how do you think the academy will receive your partiality for the sex?' She praised him ironically as 'Admirable genius!' for his talent to read the temperature when he had stated that all his thermometers had broken. When he described the Russians as fearful, she countered: 'Go! Mons. Chappe, ask the Swedes, the Prussians, the Poles, and Moustapha the Victorious . . . whether the Russians are timid!' And how, she asked, could he possibly have corrected maps, made geological observations and measured petticoats while sitting in an enclosed sledge and travelling fast as 'lightning'.

'No nation', Catherine insisted, 'has been abused with more falsity, absurdity and impertinence than the Russians'. She was not even placated by Chappe's admission that she was now forming 'a new nation' because she had commissioned several astronomers to view the transit of Venus. If anything, to Catherine's mind, with Chappe's publication of *Voyage en Sibérie*, the success of the Russian transit expeditions had become even more important. It would place her empire irrefutably at the heart of European science.

Not aware of the fury his account would provoke, Chappe congratulated himself on his publication. With *Voyage en Sibérie* fresh from the printing press, he finally left Paris in mid-September, only six days after the eighty-year-old Delisle died. Crippled by violent attacks of gout throughout the summer, Delisle had made one last appearance at the Académie at the end of August, to celebrate the king's saint's day. Having been ill, he had known for a while that he would most certainly miss the second transit – perhaps, if a Sultan's daughter had replaced astronomy as the object of his affection, he didn't care, but he certainly knew that France, once again, was contributing to the global transit project: with Le Gentil in Pondicherry and Chappe in California.

Chappe, however, was running late for his appointment with Venus. He took with him a servant, an engineer and a surveyor, a painter and a watchmaker. The Académie also decided to extend the purpose of the journey by asking Chappe to make botanical and zoological collections as well as geographical reports. In case bad weather prevented the transit observations, the expedition would at least enhance their knowledge of the natural world. As one 'intervening cloud' could 'defeat all our hopes', Chappe said, he would still be able to provide useful information about geography and natural history to 'make amends to the learned world'.

While Chappe prepared for his departure in Paris, discussions about the transit observations in the North American colonies continued at the APS in Philadelphia. At the end of September 1768, as the first snow covered the land around the Prince of Wales Fort, one of the APS members also offered to travel to Hudson Bay and suggested that the house of representatives of Pennsylvania be asked to fund the journey. When the house voted on the project, they could only agree to grant £100 for the purchase of a telescope from London. They later contributed another £100 without any conditions as to its use, but it was far from the sum needed to fund an expedition to Hudson Bay.

The members of the APS would have to concentrate on the previously agreed observations: one from the State House in Philadelphia, one from Rittenhouse's farm in Norriton and one from Cape Henlopen in Delaware. The additional £100 provided by the Assembly could be used to build an observatory in the grounds of the State House (where seven years later the American revolutionaries would declare independence) but they were 'at a loss how to furnish the Norriton observatory'. Fortunately, Maskelyne was working his magic in London and convinced Thomas Penn – the proprietor of Pennsylvania* – to assist the international endeavour. Penn obliged and immediately ordered a telescope in London for the APS. Rittenhouse, meanwhile, decided not to rely on the overworked British craftsmen, but to make his own instruments, including two telescopes, an equal altitude instrument to measure the height of the sun, and 'an excellent Time-Piece'.

Rittenhouse built an observatory in Norriton for the transit observations. Note the window in the roof which allowed him to point his telescopes into the sky.

America was growing excited about the celestial event. As advertised in the *Boston Chronicle* in late 1768, The *New-England Almanack* for the next year provided 'A particular Account of the Transit of Venus', which was illustrated with engravings of the planet's predicted path. Over the next months other newspapers across the colonies began to report on the observations and encouraged amateurs everywhere to take part. Articles explained how people should view the transit through a 'spy-glass' with smoked lenses and how to set their clocks precisely.

* The proprietors of Pennsylvania were the heirs of William Penn, who had founded the colony in the seventeenth century.

In Harvard, John Winthrop followed Franklin's and Maskelyne's suggestions and attempted to convince the government of Massachusetts to fund an expedition to Lake Superior. It was 'extreamely important to have as many observations as we can, of the whole duration of the transit', Winthrop told a member of the Council of Massachusetts. To underline the international nature of the project, he explained that many European countries were dispatching expeditions – even Catherine the Great was sending an astounding '8 companies to the northern parts of her empire'.

Winthrop also copied parts of Franklin's letters and Maskelyne's comments. It would be a 'great pity to lose so critical an opportunity', Winthrop continued. To make it easy, the expedition could be attached to the British Army, which would be sending provisions to the western forts in early spring – the astronomers could travel with the convoy 'without any great expence'. He also gave two well-attended public lectures on the transit to further advertise and emphasise its importance. Without this undertaking, Winthrop said, humankind would never be able to discover the true size of the solar system, the 'principal object' of astronomical enquiry. The lectures were published widely, together with an appendix which illustrated the best techniques for observing the transit.

In Philadelphia the members of the APS decided to form a Transit Committee to oversee their contributions, but progress was painfully slow. They were still waiting for their telescopes from London and Rittenhouse had yet to finish his observatory in Norriton because of the dismal winter weather and unreliable workmen. Following Nevil Maskelyne's suggestion of dispatching an astronomer to Lake Superior, the APS also asked the Pennsylvania Assembly for funds to send an expedition 'at least as far westward as Fort Pit [sic]'. Though Fort Pitt (today's Pittsburgh in Pennsylvania) was not as far west as Lake Superior, it was easily accessible and 300 miles further west than Philadelphia. The sun would set almost half an hour later there than on the East Coast, giving the astronomers more time for their observations. The transit, the APS Committee underlined in their petition, was of such importance that 'most of the civilised states in Europe appear to be desirous of lending assistance in it'. America's participation was essential for the 'reputation of their country' – but despite their requests no funds were forthcoming. In Harvard, Winthrop also received bad news when the governor of Massachusetts decided that he was 'unauthorised' to grant money for an expedition to Lake Superior. The colonists would have to watch the transit from towns along the East Coast instead.

As news of the favourable conditions on the North American continent spread, amateur observers in the colonies began to prepare. A wealthy merchant in Providence, Rhode Island ordered instruments from London; a gentleman from Newbury, Massachusetts asked a scientifically minded friend to observe the transit with him, as did the Surveyor General of Maryland, among many others. The British might have been the first to send an observer, the French and Spanish might have been the only ones with plans to view the transit from the West Coast of North America, but the colonists also were ready to be part of the project.

Racing to the Four Corners of the Globe

The scientific societies in the European capitals had planned expeditions to all four corners of the globe. The routes of the transit astronomers spun like invisible threads across the world, trailing into distant countries and unknown regions. Catherine the Great's astronomers left St Petersburg loaded with instruments and other provisions. Georg Moritz Lowitz had finally departed for Guryev with his young son and an entourage of seven sledges pulled by eighteen horses. In Sweden Fredrik Mallet and Anders Planman were moving north but both were encountering problems. The gloomy Mallet was trudging through three-foot-high snow and along icy rapids. It was minus 30°C and there was no warm food to be had. He regretted that he had ever chosen to become an astronomer and complained how 'ugly Lapp women are'. Planman meanwhile was in more serious trouble. He had returned to Kajana in eastern Finland to find himself in the midst of an escalating border conflict with the Russians. 'There is hardly a day or a night in which one is safe', he wrote to Wargentin. He was 'so anxious and worried', Planman said, that he slept with a 'loaded musket' next to his bed.

Alexandre-Gui Pingré had set out from France on a one-year sea voyage to test the accuracy of marine clocks, with the hope of viewing the transit in Haiti, and his compatriot Chappe was on his way to California. William Wales, at Prince of Wales Fort in Hudson Bay, woke every morning with his bedding frozen stiff against the planks at the head of his bed, which were covered with 'ice almost half as thick as themselves'. While Wales's half-pint tumbler filled with brandy turned within minutes to solid ice and his clock stopped because of the freezing temperatures, Captain Cook and his crew sailed towards the South Sea. Having organised the *Endeavour* expedition, the British ventured the farthest of all, but this was still not enough for the ambitious Nevil Maskelyne who was attempting to mount yet another expedition. Worried that the Russians and Swedes might not have sent enough observers, he had proposed (and continued to do so) several more northern expeditions to Spitsbergen, Vardø and the North Cape. Since Vardø was covered by Jesuit priest Maximilian Hell and Spitsbergen was too far away,* the Royal Society decided to dispatch a team to the North Cape. At the end of 1768, they made yet another 'application' to the Admiralty in order to procure a ship 'for carrying the Observers'. Maskelyne had the perfect team in mind: Charles Mason and Jeremiah Dixon, recently returned from their five-year surveying mission in America. To Maskelyne's surprise, though, only Dixon accepted the commission. After eight years of travelling, Mason might have been in need of rest, or of a break from his close collaborator.† Dixon on the other hand still didn't seem to enjoy being in Britain for too long and couldn't wait to embark on yet another adventure. Whatever the reasons, Maskelyne – who worried that the expedition might not come together at all – immediately put forward his assistant at the Greenwich observatory, William Bayley, as a replacement for Mason. The last voyage to the north was finally organised.

The race to chase Venus was truly underway.

---

* The tenacious Maskelyne was not quite ready to give up on Spitsbergen, a remote island which was more than 500 miles further north than the North Cape. He canvassed for the expedition at the highest level by speaking to the First Lord of the Admiralty. In the end he failed to convince the Royal Society.

† The Royal Society would eventually send Mason to Ireland where he observed the transit on his own.

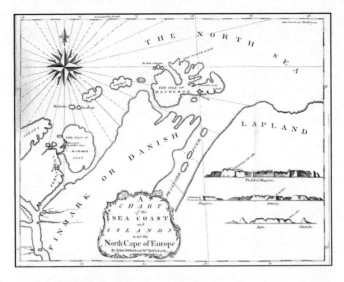

A map of the northern part of Norway with the North Cape and Hammerfest where Jeremiah Dixon and William Bayley would view the transit

 To the East: French Expedition, Le Gentil

Le Gentil was at sea once again. In May 1766, four years before the transit, he had sailed to Manila in the Philippines – only to be ordered by the Académie in Paris to observe the transit at Pondicherry instead. As always the optimistic French astronomer talked himself into seeing the positive side of his situation. Though he had built an observatory and taken all the necessary preparatory observations in Manila, Le Gentil persuaded himself to be happy about the change of plan. He would not have to deal with the capricious Spanish governor of Manila any more, a man who, Le Gentil now told himself, would have put 'obstacles' in his way. At least in Pondicherry, which the British had returned to France after the Seven Years' War, he could look forward to the cooperation of the French governor – an old acquaintance.

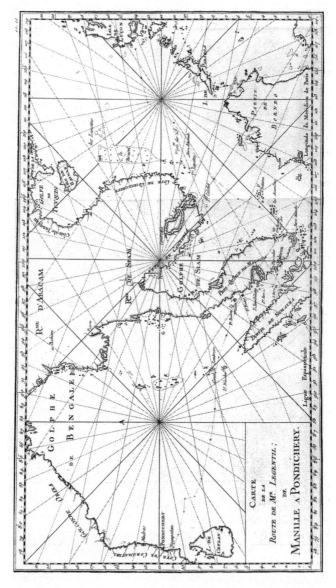

Le Gentil's map of his route from Manila to Pondicherry through the
Strait of Malacca

In early February 1768, Le Gentil left Manila on a Portuguese ship. For once everything seemed to work out for the Frenchman. They sailed through the South China Sea towards the Strait of Malacca, a narrow channel between Malaysia and Sumatra, which brought them into the Indian Ocean. The weather was glorious and Le Gentil rejoiced in the beauty of the mirrored surface of the water which, he wrote in his journal, was as smooth as a 'lake'. But just as he was praising the perfect sailing conditions, the wind picked up and the sea 'roared'. With an island to one side and a sandbank to the other, the vessel was in danger of being wrecked. To make matters worse, the captain and the first officer had such a row about their respective navigational skills that they both stormed off to their cabins – 'abandoning' the ship, Le Gentil nervously realised, 'to the pleasure of the wind'. As the ship bobbed pilotless on the waves, he dashed up and down the deck, trying to find someone who would take over. Panicking, Le Gentil himself went to the tiller: 'I took over here for the first time the office of pilot' – the many months he had spent on ships turned out to be of use, and he steered the vessel to safety.

The rest of the voyage was relatively uneventful. There were only a few storms, an unflappable Le Gentil reported. 'There couldn't be a more fortunate voyage than ours'. On 27 March 1768, precisely eight years after he had sailed from France to see the first transit, he sighted Pondicherry. It was just before six o'clock in the morning and the weather was spectacular. With more than a year to spare, Le Gentil had reached his destination with plenty of time to build his observatory and get ready for Venus.

He immediately met the French governor and went with the chief engineer of Pondicherry to find a location for his observatory, eventually deciding on the ruins of the fortress which had been destroyed by the British. The governor assisted as much as he could and 'under his auspices', Le Gentil said, 'I enjoyed at Pondicherry that sweet peace which is the support of the muses'. The strong brick walls of the fortress made for the sturdiest foundations for his observatory. No wind or storm would be able to shake his instruments. Within two months the bricklayers and carpenters had constructed a spacious building with one large central room measuring fifteen by fifteen feet and nine-foot-wide windows, which allowed him to use his large telescopes easily. He had also several smaller rooms which were his living quarters – 'I was more in touch with my work there', Le Gentil declared.

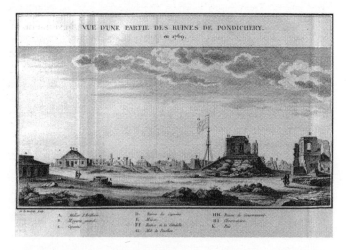

A view of Pondicherry and the ruins after the British siege. Le Gentil's observatory is the building to the right of the flagpole (labelled H).

That he was living and working on top of a vault which was also the arsenal for a staggering 60,000 pounds of gunpowder did not seem to bother Le Gentil. For a man who attracted bad luck wherever he went, he was surprisingly naïve about his explosive basement. No matter how dangerous it was, Le Gentil said, all he cared for was that he had reached Pondicherry and that the skies were clear. After years of travelling and neglect, he now had time to clean his quadrant and his clock properly. With hostilities over, the English had also been helpful, sending him a brand-new achromatic telescope from Madras. Le Gentil delighted in the clear and balmy nights in Pondicherry, so perfect for astronomical observations that they 'exceeded all expectations'. He observed a lunar eclipse in order to establish his longitude and was finally ready. His peregrination 'from ocean to ocean, from coast to coast', he declared, had been worth it – or so he thought.

To the North: Scandinavian Expedition,
Maximilian Hell

On the other side of the globe, Jesuit priest Maximilian Hell and his assistant János Sajnovics slowly moved north. They travelled to Trondheim to meet the botanist and to charter a boat to Vardø, an island at the most north-eastern tip of Norway. During the first part of the journey they were accompanied by the brother of the Danish Royal Astronomer, but he left them in Helsingör to attend to some family business.* They crossed Sweden and then drove into Norway with Sajnovics once again filling his diary with comments about food – they drank delicious hot chocolate and ate the 'most magnificent' rainbow trout and strawberries with cream but were also served 'bad wine soup'.

Hell's map of southern Sweden, showing the route from Helsingör to
Gothenburg towards Oslo

* The astronomer Peder Horrebow (brother of Danish Royal Astronomer Christian Horrebow) intended to view the transit from Tromsø in northern Norway but bad weather prevented him from getting there. In the end, he went to Dønnes – north of Trondheim – but didn't see anything.

The further north they ventured, the more harrowing the journey became. Like Chappe on his journey to Siberia for the first transit, they had great difficulties transporting their heavy equipment on the rough roads. As they crossed the mountains, they had to find ten additional horses to pull the carts up the slopes – only to struggle to maintain control on the rapid descents on their way down. Large rocks in the roads hindered their progress, an axle broke, and their progress slowed to a crawl. At some point the roads simply ended and they had to navigate through fields and pasture. By the time they reached Oslo (then called Christiania) the damage to the coaches was so serious that they bought new ones, distributing their heavy luggage on to five sturdy carts. News of their strange quest to observe the transit of Venus quickly spread throughout the city and as they left 'half the population', Sajnovics noted, 'ran after us for half an hour'.

On their way to Trondheim they saw snow-capped mountains – the first reminder of the icy landscape that would soon be their home. The views might have been awe-inspiring but the roads were becoming worse. Once again Sajnovics moaned about the food. With no taverns or inns, they had to purchase 'bad' meals from the little farms, he complained, but they quickly discovered that the local priests were more 'hospitable'. Sometimes they couldn't find anything and went to bed hungry, which was not too awful for the ascetic Hell but far more problematic for his gourmand assistant.

As the countryside grew more mountainous and isolated, accidents became more frequent. It rained incessantly and they couldn't see the narrow paths at all. Wheels and axles broke regularly, and they had to navigate across precariously small and often rotten bridges. Steering blind through thick fog, the astronomers and their carriages were in constant danger of falling into deep precipices besides the paths. For much of this part of the journey Hell and Sajnovics walked behind their coach.

When they finally reached Trondheim they were exhausted, but the most taxing part of their expedition still lay ahead. The rest of their voyage to Vardø would be by ship along the treacherous coast. With autumn coming, the winds were picking up and storms were predicted. Sailing out at sea would be rough and Hell decided instead to take the longer route along the jagged inlets of the western coast of Norway, where they could anchor in the small protected harbours. They still had a long cold journey before them.

Hell's ship sailing past the island of Torgatten on the way to Vardø

At the end of August 1768, Hell and his small team sailed for the Arctic Circle. They were accompanied by five enormous whales which staged a majestic diving display in the grey water. Once in a while they spouted huge fountains through their blowholes. Hell watched the fascinating spectacle by the light of the long northern summer night. For the next six weeks their vessel performed an erratic dance on the rough sea. Icy waves crashed into their small cabin and several times they feared for their lives and instruments, worried that the sea would 'bury' them forever. Only the sailors, they observed with disbelief, became increasingly happier, 'the wilder it roared around us'.

On 11 October 1768, they arrived at the Danish garrison in Vardø. Like Le Gentil in Pondicherry, they immediately searched for a location for their observatory. They settled on the centre of the small town but with hardly any timber on the island, Vardø's commander advised them to send their boat to the mainland to procure the necessary building materials. Amid permanent darkness, snowstorms and freezing temperatures, it took two months to build the observatory. The lack of timber and the dismal weather delayed them but not as much as the lazy carpenters, Hell complained.

The observatory in Vardø. Hell's and Sajnovics's living quarters were to the right and the observatory room, with hatches in the roof, was to the left.

When the observatory was completed in mid-December, Hell and Sajnovics set up their instruments. Several hatches in the roof and walls allowed them to take observations in order to establish their longitude but they were still hampered by the bad weather. Hell battled with almost continual fog and cold. He couldn't see any stars, the provisions froze and burst their containers, and with an icy wind blowing through the shutters and walls, sleeping was almost impossible.

Conditions deteriorated, and Hell began to fear that their little hut would be carried away by the vicious storms that whipped across the island. The snow reached the roofs of the houses, and they had to light candles all day long. Sajnovics continued his laments about the food, telling the cook that he couldn't endure the 'dry, unpalatable Norwegian dishes any more'. But despite the hardship, they also relished the magical moments – the Northern Lights illuminating the sky in kaleidoscopic swirls. With a long winter ahead the astronomers had not much to do but to watch the sky, shoot birds and seals, and wait for Venus to march across the sun on 3 June.

To the South: British Expedition, James Cook and the Endeavour

While the lives of the astronomers in the north came to a frozen standstill, the *Endeavour* sailed into the heat. After some initial seasickness, the men quickly adapted to the rolling rhythm of the waves. Most of the time the wealthy botanist Joseph Banks and his entourage of naturalists and painters entertained the rest of the crew with their curious obsessions with strange fish and seaweed, which they caught with nets. One morning a flying fish even flew through the porthole into Charles Green's cabin. The astronomer diligently delivered the specimen to Banks and returned to his own studies. Throughout these early weeks of the voyage, Green taught James Cook and some of the officers Maskelyne's lunar method to determine their longitude. He was 'Indefatigable in making and calculating these observations', Cook noted.

As they neared the equator, the crew grew excited. Every day Green saw the sun climb a little higher at noon, the sign that they would soon sail into the southern hemisphere. Temperatures rose and the air was heavy with moisture. Leather-bound books were covered in white mould, razors became useless and knives rusted in the sailors' pockets. On 25 October, exactly two months after they had left Plymouth, the *Endeavour* crossed the equator. The crew conducted the infamous 'ducking' ceremony for those who had never crossed the line – the equinoctial virgins were tied to a contraption of wood and rope, and then lowered over the side to be dipped three times into the swirling water.

On 14 November 1768, they reached Rio de Janeiro (then in the hands of the Portuguese) with the distinctive outline of Sugar Loaf Mountain rising from the shore. Cook wanted to fill their stores with fresh supplies for the next part of their journey and as required went to the Portuguese viceroy to deal with the necessary formalities. To his surprise his hosts were not very welcoming: the crew was forced to remain on board, with Portuguese soldiers rowing around the *Endeavour* to ensure that no one could leave.

Only Cook – accompanied by an armed Portuguese guard – was permitted to go ashore and buy much-needed food and water. He tried to explain that they were on a scientific mission but, he wrote despairingly in his logbook, the Portuguese viceroy 'certainly did not believe a word about our being bound to the South ward to observe the transit of Venus.' It was obvious, the viceroy explained to Cook, that they were either spies, smugglers or merchants. The transit observation was clearly 'an invented story' to cover up the real purpose of the voyage. The viceroy was unable to understand the importance of the transit and thought the idea was as absurd as the 'North Star Passing thro. the South Pole'.

Being virtually imprisoned on the *Endeavour*, Cook forced his crew into a furious regime of cleaning and repair, despite the unbearable heat of the South American summer. The *Endeavour* was 'heeled down' and, Banks's botanist complained in a letter to the Royal Society, 'we hardly cou[l]d walk'. Sailors were climbing up and down to redo the rigging. The ship was freshly caulked, the sails stitched, and every nook and cranny scrubbed. Everybody was moaning but Cook was relentless, working his frustration into the timbers of the *Endeavour*.

Banks also was growing impatient. As a wealthy landowner he was not used to ill-treatment. 'I am a Gentleman, and one of fortune', he wrote indignantly to the viceroy. He had spent an enormous amount of money to join the *Endeavour* voyage and was risking his life in order to collect plants. Peering through his telescope, he saw humming birds hovering over the strange blossoms and exotic trees laden with fruit. So close, yet so far away. He felt, Banks wrote to a friend, like a 'French man laying swaddled in linnen between two of his Mistresses, both naked [and] using every possible means to excite desire'.

Every day Banks and Cook bombarded the viceroy with letters, first pleading and then becoming increasingly angry and frustrated. As Cook worked himself into his cleaning frenzy, Banks 'cursd, swore, ravd, stampd', pacing up and down the deck. But to no avail. The viceroy thought 'it impossible that the King of England could be such a fool as to fitt out a ship merely to observe the transit of Venus'.

Annoyed but with their stores replenished, they left Rio de Janeiro in early December. Three weeks later, they celebrated Christmas with an extra allocation of rum. While Hell and Sajnovics treated themselves to a chaste hot chocolate on 25

December in Vardø, Cook's men enjoyed a more bawdy feast where 'all hands g[o]t abominably drunk', Banks reported. But the hot weather soon changed, and temperatures dropped rapidly. At Cape Horn, which was feared for its volatile winds and vicious currents, the *Endeavour* was tossed by icy gales. Waves buffeted the vessel so hard that the furniture overturned and Banks's entire library tumbled through the cabin. During the nights they were bashed in their hammocks against the ceiling and walls. On an ill-fated plant-collecting expedition to Tierra del Fuego at the southern tip of South America, disaster struck when some of the men were caught on shore by a sudden blizzard and perished.

The *Endeavour* at Matavai Bay

As they sailed into the unknown emptiness of the South Pacific, Cook amazingly succeeded in finding the island that Captain Samuel Wallis had described the previous summer. Eight months after leaving England, on 13 April 1769, the *Endeavour* arrived in Tahiti.* From the distance they saw

---

* The *Endeavour* was only the third European vessel to arrive in Tahiti – the *Dolphin* under the command of Captain Samuel Wallis had been the first in June 1767, closely followed by the French captain Louis-Antoine de Bougainville in April 1768.

the peaks of the mountains rising from the sea, and when they anchored at Matavai Bay which was ribboned by black beaches, the Tahitians greeted them peacefully. The *Endeavour* crew walked in the shade of palms and trees laden with coconuts and breadfruits. They called it 'the truest picture of an arcadia of which we were going to be kings'.

Cook immediately ordered fifty of his men to dig trenches and ramparts as well as to fell trees in order to built a fort 'for the Defence of the Observatory' at the northern end of Matavai Bay. One side of the fort was bordered by the river while the other sides were protected by a high fence. Inside the compound were several tents: one that functioned as the observatory (topped with the British flag), one for the kitchen and another for the guards, amongst several others. Cook also installed some swivel guns and cannons. They called it 'Fort Venus' in honour of their mission. On the palm-fringed island dotted with groves of trees and small open huts, the European fortifications that enclosed the fort must have looked completely out of place.

Fort Venus. The observatory tent is the round structure in the middle with the flag.

Despite the construction work, the men still found time to enjoy this bucolic idyll, and sampled the pleasures of free love and 'lusty' females. The island's women unequivocally showed their willingness to share their sleeping mats with the sailors, though Banks complained that the huts were open so that they couldn't put 'their politeness to every test'.

On 28 April, coincidentally but with perfect symmetry, Cook ordered the prefabricated observatory to be carried from the *Endeavour* to the fort, just as on the other side of the globe Maskelyne's former assistant, William Bayley, erected his own on the barren rock of the North Cape.* Three days later, on 1 May, Cook and astronomer Charles Green began to bring their instruments on shore, but realised, the next morning, that a large quadrant was missing. With the guard only five yards away, it seemed impossible that someone could have stolen it during the night – but the instrument had vanished.

They peered into every corner of Fort Venus and on board the *Endeavour*. Without the quadrant, the expedition would be worthless. Green would not be able to measure the altitude of the sun, which was needed to set the clock (and to calculate the longitude). He would also not be able to take the altitude of planets, moons and satellites necessary to determine their latitude. The observation of the transit of Venus was only useful in combination with the exact geographical position of the observer – the quadrant was essential to the success of the voyage.

One of the Tahitians must have taken it, Cook fumed. It was not the first time that they had stolen. They took 'every thing that was loose about the ship', from knives, snuffboxes to even the glass panes from the portholes. In this, Cook said, they were 'prodiges expert'. The Tahitians believed, so Banks observed, that once they had taken something 'it immediately becomes their own'.

Cook had not sailed across the globe through storms and dangerous seas to fail now. They had endured too much to be defeated by a thief. With the quadrant missing, Cook discarded his own command to treat the Tahitians 'with all imaginable

---

* On the same day Hell's expedition in Vardø almost came to a premature end when some stray pellets (a few young locals were shooting birds nearby) went through one of the observatory windows, missing him narrowly.

humanity', and ordered that some of the 'Principle people' be locked up until the precious instrument was found. As Cook began to question his prisoners, Green together with Banks and the midshipman scoured the island. Banks, who had been the chief trader since their arrival, made use of his contacts and discovered that one of his new Tahitian acquaintances knew the culprit. They rushed from one hut to another, questioning and cajoling. By late afternoon and seven miles from the fort, they finally discovered the person who had taken their quadrant.

After some discussion (and a display of the pistols that they had brought), Banks and Green persuaded the Tahitians to return the stolen goods. Green watched in horror as they brought out one part after another: the islanders had dismantled the delicate instrument to divide it between them. But after examining the pieces, Green was relieved to see that there seemed to be only minor damage. They packed the quadrant in a box and padded it with grass before beginning their long walk home. By the time they returned to Fort Venus it was dark. It had been a long, hot and tiring day – they were exhausted but triumphant. Everything they needed for the transit observations was in place. They had only to guard their instruments carefully for another month.

Dancing and music in Tahiti, including a man playing a flute with his nostrils

To the West: French Expedition,
Chappe d'Auteroche

In mid-September 1768, Chappe d'Auteroche left Paris for California. It was not to be an easy journey. The stormy passage from Le Havre in France to Spain took twice as long as expected and upon arrival in Cadiz, Chappe endured a long bureaucratic tug of war with the Spanish authorities who had failed to provide passports for his assistants. Without their papers, they weren't even allowed to step on the deck of the Spanish ship that was meant to take them to Veracruz in Mexico. Even worse, Chappe was told he was only allowed to take one instrument, an order that would have made the transit observation impossible.

The main part of the long journey still lay ahead: the ocean crossing from Cadiz to Veracruz on the eastern coast of Mexico, then almost 800 miles on horseback across country to the town of San Blas, and finally the second sea voyage up to Baja California, the 500-mile-long headland between the Pacific Ocean and the Gulf of California. There had already been so many delays, Chappe wrote in his journal with typical flourish, 'which a thousand times made me despair of getting in time to California'.

He begged the governor of Cadiz to help, only to hear that he would have to apply directly to the Spanish Court in Madrid. Couriers were dispatched, and for weeks Chappe waited impatiently. Though the king of Spain had granted him permission to view the transit of Venus from his territories in America, there were still too many formalities involved. Time was running out. 'If we were retarded ever so little longer', Chappe complained, it would be 'morally impossible' to get to California for the transit. Making use of every official connection and acquaintance he had, Chappe wrote letters to apply for additional passports and requested permission to embark on the first ship – 'no matter which' – to sail for Veracruz. To his surprise the Court in Madrid suddenly agreed and orders were issued to 'instantly fit out' a small vessel for the purpose. Once the instruments were put on board, they were ready to depart. It was now 21 December 1768, two long months after Chappe had arrived.

The ship was tiny – Chappe, his entourage and the Spanish observers Vicente de Doz and Salvador de Medina, were crossing the rough winter Atlantic in a boat with a crew of only twelve. Chappe felt the light vessel bobbing up and down like a 'little nut-shell', but he did not care. No matter that he had been warned about the 'frailty of the vessel', he declared exuberantly that it would be to their advantage – its size made it swift and in his eyes better than 'the finest ship' in the entire Spanish fleet. As Cadiz disappeared at the horizon and the wind carried them towards their destination, Chappe felt 'at that instant a transport of joy'.

Despite his initial glee, the ennui of the sea journey soon proved too much for him. During the long days, the jovial Frenchman entertained himself by calculating the longitude according to Maskelyne's lunar method, though he found the calculations 'tedious'. Maskelyne might have liked the rhythmic repetition and order but Chappe, with his penchant for exaggeration and adventure, was easily bored. Sea voyages, he moaned, were just too 'tiresome and uniform'. Where other astronomers were happy to concentrate on their observations, he preferred to dip into the colourful world of his imagination. With no adventures or pirate attacks to report, he entertained himself by dreaming up 'a thousand times' Christopher Columbus and his audacious explorations. Chappe was counting the days until he had solid ground under his feet again.

It took seventy-seven days to reach Veracruz on the east coast of Mexico. Relieved to have survived the sea voyage, Chappe's initial euphoria turned quickly to panic when Veracruz was hit by a three-day hurricane. Cut off from his precious luggage, which was still on board ship, Chappe waited on shore 'in greatest anxiety for my instruments'. If they were damaged, he had travelled in vain across half the globe.

When it was calm again, and his telescopes and clocks were safe, Chappe organised the next leg of their journey. They still had to cross the mountainous country to reach San Blas at the western coast of Mexico and then sail to Baja California.*

---

* The first part of the journey was along the same route that the Spanish conquistador Hernán Cortés had taken exactly 250 years previously when he had led the expedition that destroyed the Aztec empire.

Everything was repacked into small bundles so that the heavy instruments could be carried by mules. Their progress through a region that was known for its murderous 'banditti' was painfully slow. The roads were 'frightful', the heat 'excessive' and the mules were hardly moving at all. They rode along mountain paths so narrow that the larger instruments, tied to the pack animals, dangled precariously over precipices. The local food was appalling, Chappe complained, so hot and spicy that it was inedible 'especially for a Frenchman'. Once again he entertained himself by taking notes on the women who, in one village, were half-naked and displayed 'a most frightful neck'. The women in Mexico, the proficient expert concluded, were 'no very pleasing figures'. He was back in his element.

Chappe was on horseback – choosing speed over comfort – but his Spanish colleagues delayed the journey by travelling in style in a carriage. No matter how fast Chappe was, he had to wait for them. By King Carlos III's orders they had to watch his every step. Under no circumstances, so the king's instructions to Doz and Medina read, were they to lose sight of the French astronomer. The atmosphere was tense and they did not talk much – the French team under Chappe kept to themselves most of the time.

They eventually reached San Blas on 15 April, but the Spanish continued to slow them down. Whereas Chappe was ready to sail immediately, Doz and Medina wasted time by collecting the timber they needed to construct a large observatory. He only took enough material, Chappe noted indignantly, 'to make a tent, and a great beam of cedar on which to hang up my clock'.

The problems did not abate. The passage from San Blas to Baja California was hampered alternately by contrary winds or lulls. They were so much slower than expected that they ran out of food and water. Chappe 'began to despair' and was preparing himself for the 'most cruel disappointment'. By mid-May 1769, when the other astronomers from the Arctic Circle to the South Sea, from India to the distant corners of the Russian empire, were all installed in their temporary observatories, Chappe still had not arrived. But he was determined to land at the first place they could reach.

'I little cared whether it was inhabited or desert, so as I

could but make my observation', he insisted, but his Spanish colleagues refused to disembark. Chappe paced up and down the deck of the small vessel. He was furious, arguing for his life, his career and the future of astronomy. Admittedly, the impatient astronomer conceded, it was not the best place to land. The swell was strong and the winds contrary. The Spanish feared that the vessel would be smashed and suggested continuing another forty miles to reach a safer harbour. This, Chappe argued passionately, would take several days. They would have to tack against the wind and currents. In any case, he was sure that the king of Spain would 'rather lose a poor pitiful vessel, than the fruits of so important an expedition'. Chappe was not going to give up so close to the destination. Appealing to the Spaniards' manliness and pride, he tried one more time, insinuating that others had successfully landed here before.

After hours of refusing to accept any other option, he convinced his colleagues. For the moment Chappe's fury seemed to eclipse the dangers of the angry sea. Once the decision was made, the captain turned the vessel and sailed towards the coast. As they came closer, a fresh gale drove them hard towards the rocks and the Spanish regretted having listened to Chappe, blaming him for their impending deaths.

The only way to land was to get the instruments and luggage ashore in their dinghy. The small open boat bobbed on the surf, waves crashing over them. They were soaked, as were their trunks. Clothes could be dried, some of the instruments too, but if the pendulum clock or the telescopes were to get wet, Chappe would not be able to measure Venus's path. Without the instruments, it wouldn't matter if he arrived on time or not. When it was Chappe's turn to leave the ship, he wrapped up his precious clock and sat on it 'to keep it dry'. An endless onslaught of capricious waves rolled over the dinghy, the surface of the water a swirling dance of white foam. He heard the 'horrid roaring' as the surf crashed on to the shore and the rocks. The sailors rowed with 'all their might' and Chappe watched their strong arms battling against the power of the sea, certain that they would not make it.

Day of Transit, 3 June 1769

The day of the second transit had finally arrived. Across the world astronomers waited for Venus to appear. Le Gentil had lived in Pondicherry for more than a year, preparing for the big day; William Wales had been in Hudson Bay since the previous summer; Cook and the *Endeavour* crew were in Tahiti; Maskelyne's assistant William Bayley had landed at the North Cape on 28 April and Jeremiah Dixon nearby in Hammerfest on 7 May.* With more than eighty observers at thirty viewing stations in Britain and sixteen abroad (not counting the North American colonies), the British were clearly in the lead, followed by the French with almost fifty astronomers at eighteen locations in France and five overseas.

There were astronomers in nine German towns, and Dutch observers positioned in Leiden. The Swedes were also prepared. Pehr Willhelm Wargentin had recruited twenty-one observers at nine locations in Sweden and Lapland. Anders Planman was safely installed in his observatory in Kajana and Fredrik Mallet – despite the fierce winter conditions and his bad mood – had made it to Pello on 12 May, late, but in time.

Eighteen astronomers were stationed at ten locations on Russian soil. Here the astronomers from Germany played leading roles: one arrived in Orenburg in mid-March, and

---

* It had been decided that Bayley and Dixon should split once they reached the Arctic Circle to increase the chances of at least one successful observation. Dixon's arrival had caused much confusion. When his frigate anchored in Hammerfest, the inhabitants feared that the British had come with military intent, bringing 'destruction into their country' – but Dixon convinced them of his peaceful mission.

another had reached Orsk where he had taken his first astronomical observation on 9 April. Georg Moritz Lowitz, the astronomer from Göttingen who was travelling with his young son to Guryev on the Caspian Sea, only just reached his destination, racing against the thaw, crossing rivers on precariously thin and floating ice sheets. Two weeks before the transit, Lowitz had decided that his entourage and instruments slowed him down too much and sped ahead to build the observatory while his team followed at a slower pace. On that same day, the German astronomer and Jesuit priest Christian Mayer entered, for the first time, the observatory in St Petersburg where he would view the transit. Recommended to the Russian Academy by Jérôme Lalande, Mayer had only left his home in Mannheim on 3 March, rushing to reach St Petersburg.* The Swiss astronomers from Geneva, as well as the Russian Stepan Rumovsky had also reached their respective viewing stations on the Kola peninsula. Even the observer who had gone all the way to Yakutsk had arrived in time for the transit.

On the North American continent forty-seven observers anticipated Venus's appearance – from the local administrator of a mine near Mexico City to the 'Surveyor-General of Lands for the Northern District' in Quebec. More than thirty observers had spread out over twelve locations along the East Coast, including John Winthrop in Cambridge. Benjamin Franklin had finally 'after much Delay and Difficulty' dispatched Winthrop's instruments from London in mid-March and the American Philosophical Society in Philadelphia had received their telescopes a few days before the transit.

Even Chappe and his team had made it. When they had arrived on the southern tip of Baja California on 19 May, there had been little hope for a successful transit observation. Miraculously they had survived their dangerous landing with not a single instrument damaged. They found shelter in the nearby mission of San José del Cabo, a community that was

---

* Mayer had taken a huge detour via Amsterdam to pick up the instruments which he had ordered from England. Lalande had told Mayer to go to Russia but had forgotten to tell the Academy in St Petersburg. When a Dutch newspaper reported that Mayer was travelling there, the members of the Russian Academy were rather surprised and thought Lalande's behaviour 'a bit strange' but were nonetheless delighted.

ravaged by the outbreak of an 'epidemical distemper' – typhus, which had already killed one-third of the population. Unsurprisingly, the Spanish observers feared for their lives and suggested continuing the journey overland, but there was no arguing with Chappe. He would rather risk his life than miss the transit. He would 'not stir from San-Joseph', he declared, 'let the consequence be what it would'. He was ready for Venus.

On 3 June 1769, as one location after another emerged from night to day, astronomers and amateurs across the world braced themselves. It was the last transit that any of them would ever be able to watch.

 The South: British Expedition, James Cook and the Endeavour, *Tahiti*

The *Endeavour* crew was nervous. Since their arrival in Tahiti in mid-April, the sky had been overcast for much of the time, which 'makes us all not a little anxious for success', they worried. As the day of the transit approached, the weather improved, but there were still too many clouds in the sky. James Cook decided to dispatch two teams to neighbouring islands to make additional observations, 'for fear that we should fail here'.

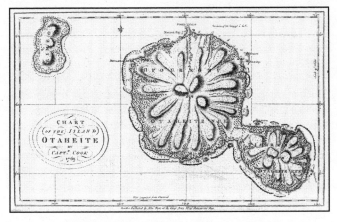

A Map of Tahiti showing 'Point Venus' at the top

In the days before the transit, Cook and the astronomer Charles Green had been 'very buisy' preparing their instruments and instructing the teams that were to observe the transit on the other islands. Telescopes were tested, lenses polished and the clocks checked one final time. On 1 June one team, and on 2 June the second, left Matavai Bay, their boats loaded with equipment and their heads full of Green's instructions. At Fort Venus an apprehensive Green continued to fuss over his own instruments. As the anticipation built, the men worked silently next to each other, 'all Hands anxious for Tomorrow'.

Then, as the sun rose on 3 June, Cook and his crew woke to a clear sky. They could hardly believe their luck – there was not a single cloud to be seen. 'The day', Cook wrote in his journal, 'prov'd as favourable as we could wish'. As he, Green and the *Endeavour*'s botanist Daniel Solander (who had been asked to man the third telescope in Fort Venus) settled at their instruments, they could do nothing more than wait for Venus to push on to the disc of the sun. Cook had ordered sentinels to guard the fort so that none of the Tahitians 'might disturb the Observation'. With each passing minute the tension rose. No one uttered a word.

Green was the first to see something – at 9.21 a.m. and forty-five seconds he noted a light at the edge of the sun. Five seconds later Cook detected the 'first visible appearance of ♀',* but it took Green another ten seconds before he was certain that he was really seeing Venus. Solander was still not sure. Like the astronomers during the first transit, the three men were struggling to determine the exact moment of entry. Solander noted a 'wavering haze' and Cook some 'undulating'. The exact timing, Cook wrote, was 'very difficult to judge'.

Undeterred, Cook and Green continued their observations. During these hours the temperature climbed to a punishing 48°C. The heat became 'intolerable', the men complained. Then, just after three o'clock in the afternoon, with the sky still brushed clean off any clouds and not a hint of a breeze, they awaited the exit. As Venus and the sun slowly separated, Cook and Green noted the times – but the two astronomers were twelve seconds apart. Venus had lingered, 'which of course', Cook wrote, made the timing 'a little doubtful'.

* The symbol ♀ was the sign for Venus.

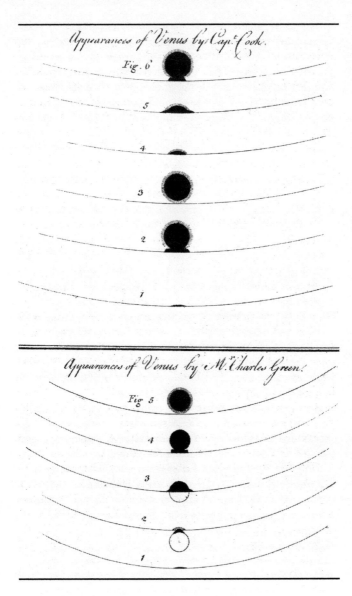

Drawings made by James Cook and Charles Green, illustrating the
luminous ring around Venus and the black drop effect as seen in Tahiti

The discrepancies between Solander's times and theirs could be explained by the different magnifying powers of the telescopes – but Green and Cook had used exactly the same model. The other two teams had encountered similar problems, they reported on their return, but despite these setbacks they all hoped that the Royal Society would be satisfied with the results. The weather had been perfect and Venus's hesitation was not their fault. They had done all they could.*

 The West: French Expedition, Chappe d'Auteroche, San José del Cabo, Baja California

In California too the sun graced a blue sky. Since their arrival two weeks previously, Chappe d'Auteroche and his team had worked at a ferocious pace in order to be prepared. 'I had still time enough', he had reassured himself. Realising that he could make it, he 'felt such a torrent of joy and satisfaction', he declared with his usual panache, 'it is impossible to express'. A large barn in the mission of San José del Cabo became his temporary observatory (Doz and Medina built their own). Chappe had hurriedly ordered half the roof to be taken off so that they could point their telescopes towards the sky. After their delayed arrival, he had set up his instruments 'as they were', because there had been no time to adjust them – but he was thankful that they had made it at all.

During the days before the transit, as the astronomers prepared their makeshift observatories, the people of San José succumbed to typhus at an alarming rate. As Cook's crew listened to Tahitian music in the South Pacific, Chappe heard nothing but the 'groans' of the stricken inhabitants. But with the determination of a man who was prepared to die in the pursuit of knowledge, he 'cared for nothing else' but the transit. No matter what, Chappe would observe Venus's path.

---

* Meanwhile other members of the *Endeavour* crew were less interested in the astronomical nature of their expedition. Some used the moment of the transit to raid the stores. Botanist Joseph Banks, who had accompanied one of the observer teams to the neighbouring island, went botanising (as well as showing the Tahitians that the astronomers were interested in a small dot that was crossing the sun). He celebrated the success by seducing three 'handsome girls' who, he was delighted to discover, 'with little perswasion' agreed to sleep in his tent.

When he woke on 3 June, 'the weather', he said, 'favoured me to my utmost wish'. They were expecting Venus at noon, and so he spent the morning making last-minute preparations. During their ocean crossing from Cadiz to Mexico, he had written instructions for the transit day. He pinned the long list to the wall – 'so that I might at every moment be able to recall what I was to do or to prepare for'. He had brought the latest achromatic telescopes from London and the clock was fixed to a cedar block buried two feet in the ground to make it as stable as possible. With pedantic precision Chappe had also enclosed the clock in a box, which he then covered with paper to protect the delicate mechanism from wind and dust. Everyone knew their role: Chappe was observing, his servant was to count the minutes and seconds, the engineer was ordered to record the times and the watchmaker assisted with the instruments.

Then, only seconds before noon, Venus slowly moved on to the edge of the sun. Like Cook and Green, Chappe noticed that Venus seemed to be stuck for a moment to the sun's edge, 'detaching itself with difficulty'. The black drop effect which astronomers had detected in 1761 complicated the timings once again. For the next hours Chappe measured and observed. Everything worked according to plan. He timed the internal exit at 5.54 p.m. and fifty seconds, and eighteen minutes later the external exit – then the black dot finally disappeared and the transit was over. Looking at his long list of times and measurements, Chappe couldn't believe his luck. 'I had an opportunity of making a most complete observation', he recorded.

It was to be the last sentence he wrote in his journal.

 The North: Scandinavian Expedition, Maximilian Hell, Vardø, Arctic Circle

Vardø was hidden under a thick blanket of white. A few weeks previously yet another storm had bucketed more snow on to the little island, but at least the long months of winter darkness had ended. Maximilian Hell and his assistant János Sajnovics had lived a secluded life since their arrival in mid-October. Only once had letters from Copenhagen been delivered, but strangely they were better informed about the progress of some of their fellow astronomers than their colleagues in the scientific societies of London or Paris. Only three weeks before the transit, a Norwegian captain had brought news that Jeremiah Dixon and William Bayley were preparing for the transit in Hammerfest and on the North Cape. Two days later, on 14 May, the crew from another ship told them that one of the Russian observers had died on Kildin, a small island north of the Kola peninsula in the Barents Sea.*

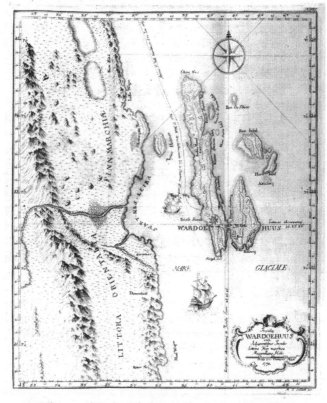

Hell's map of Vardø with the exact longitude and latitude of the observatory

* The Russian astronomer Ochtenski was the first transit astronomer to die for Venus, but nothing is known about the circumstances of his death. He was assistant to Stepan Rumovsky, who had dispatched him to Kildin.

Saddened to hear about the death of one of their fellow transit astronomers but hopeful that Dixon and Bayley would succeed, Hell and Sajnovics turned to their own preparations. On 2 June they carried out one final check of their instruments. They were ready for the 'great observation' but too nervous to sleep. Though Venus would only appear after nine o'clock in the evening, they woke in the early hours of the morning. Pushing aside the shutters from their windows, they could see the sun clearly. It was a promising start, but by no means a guarantee that good weather would continue. During the summer months in the Arctic Circle, thick fog regularly rolled in from the sea. Anybody who lived in Vardø would have been able to tell Hell that the likelihood of a continuously clear day on 3 June was relatively small.

Sure enough, minutes later the sky was overcast, but an hour later the astronomers were bathed in bright sunlight. Throughout the day the clouds played a tantalising game of hide and seek with the sun. At 3 p.m. the sky was covered in white clouds, at 6 p.m. the sun briefly peeped through. At 9 p.m., with the transit only minutes away, Hell and Sajnovics pointed their telescope in readiness. Then, just as Venus was predicted to appear, the clouds opened: 'with the special grace of God', the astronomers exclaimed, as they saw the small black dot. The commander of the garrison immediately raised the flag and Vardø's inhabitants came running to the observatory in order to catch a glimpse of the celestial encounter. But by the time they had assembled at the telescope, the sun had disappeared again.

For six long hours, throughout the light northern night, Hell and Sajnovics watched in hope, stunned that the clouds refused to free the sun from their dark embrace. 'Unbelievable!', Sajnovics wrote in his diary, 'but nevertheless true'. It had taken them almost six months to get from Vienna to Vardø, and during their seven and a half months in the Arctic Circle, they had endured treacherous storms, ice, snow and a winter of seemingly never-ending darkness – only to be defeated by clouds. There was absolutely no chance, everybody agreed, that they would see Venus again.

Bored from staring at a grey sky, most of Hell's audience went to bed. Then, at three o'clock in the morning of 4 June, as Venus like a shy maiden prepared for her exit behind a cloudy veil, the wind picked up and the clouds dispersed. At once Hell and Sajnovics could see the black dot slowly moving towards the edge of the sun. They couldn't believe their fortune and carefully noted the times of Venus's exit.

The merchant of the little town who had remained in the observatory was so excited that he fired three small cannons in celebration, rupturing the silence. After a successful observation, the devout Hell sang the hymn 'Te Deum laudamus' to thank God for his mercy and then went to rest. It had been a good day.

 The East: French Expedition, Le Gentil, Pondicherry

In Pondicherry the beginning of the transit would happen in the darkness of the night. As the sun rose over the Indian Ocean on the morning of 4 June, Venus would already have begun her march. The prospects for a successful observation were excellent. For more than a month, the blue dome of the morning sky had shown not a trace of cloud. On the evening before, Le Gentil and the French governor of Pondicherry had seen Jupiter's satellites clearly. His acquaintances and neighbours already began to 'wish me luck', Le Gentil noted in his journal. Tomorrow, the French astronomer was certain, would be the perfect day to watch Venus glide across the face of the sun. One last look into the night sky confirmed that there were still no clouds. For nine long years Le Gentil had waited for this moment. It was his final chance to leave a legacy as an astronomer. 'With my soul content', he said, he awaited the auspicious day 'with tranquillity'.

At 2 a.m., Le Gentil was woken by the 'moaning' of the sandbanks and hurried to the window. The sky which for the previous months had been brightly illuminated by stars every night was covered by clouds. It was entirely calm, and with no wind, there was no hope that they would disperse. 'From that moment on, I felt doomed', Le Gentil said. As he lay on his bed with his eyes open, unable to sleep, he listened. At five o'clock in the morning, he heard the wind blowing 'ever so little' which gave him a glimmer of hope, but within a few minutes the weather turned and the breeze had become a squall. The sea was suddenly laced with dancing foam and the air thick with whirling sand and dust. With renewed fury the wind brought more clouds which spread out, as Le Gentil despairingly noted, to form 'a second curtain' that hid the rising sun.

At 6 a.m. the storm died down but the clouds lingered. Le Gentil could not see even a trace of the sun. An hour later, just as Venus prepared for her final exit, he could detect only a 'light whiteness' that radiated from behind the clouds. Instead of the fiery ball of the sun, pinpricked by the moving dot of Venus, Le Gentil saw absolutely nothing of the transit.

By 7 a.m. the transit had passed unseen by the French astronomer. Half an hour later, as if the heavens were mocking him, the sun burned down upon his face. 'I had difficulty in realising that the transit of Venus was finally over', he said. It seemed as if the clouds had appeared purely to cause him 'chargrin'. For the past nine years he had travelled tens of thousands of miles, crossed the oceans and risked his life several times – 'only to be the spectator of a fatal cloud'.

*　　*　　*

Some 250 observers at 130 locations had aimed their telescopes at the sky. In Europe many watched, even if it was only for a few minutes, before darkness hid the sun. In Philadelphia, the APS had succeeded in mounting their three observations. American astronomer David Rittenhouse woke early on the transit day to see the sky gleaming with a 'purity of atmosphere'. A large group of Philadelphians had arrived at his farm in Norriton to watch the transit. Then, just after two o'clock in the afternoon, as Venus readied herself, Rittenhouse became so over-excited that he collapsed and fainted, missing the beginning of the most important event of his scientific life. When he regained consciousness, he quickly grabbed his telescope to discover that Venus had already entered the sun, but calmed himself sufficiently to take some observations.

In Russia, at a country estate thirty miles from St Petersburg, Catherine the Great had invited eighteen of her favourite courtiers and the German astronomer Franz Aepinus to view Venus's exit just after 3 a.m. She played cards all night 'without any rest' so that she would not fall asleep and miss the event. In Britain King George III observed the beginning of the transit with his wife and four astronomers from his brand-new observatory in the Old Deer Park in Richmond, before nightfall erased the sight.

There were observers everywhere: missionaries in China, Charles Mason in Ireland, and employees of the East India Company in Madras. In Jakarta a wealthy Dutch priest had built a lavish six-storey, eighty-foot high observatory (at the cost of more than double that of the governor's palace), which would become one of the most famous sights in the East Indies.* Even two amateurs in the Philippines, who had been trained by Le Gentil when he stayed in Manila, succeeded in following Venus's slow march.

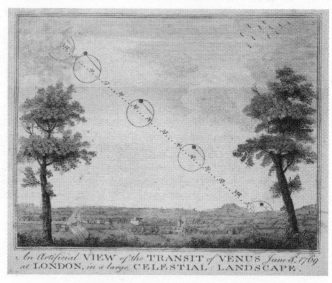

An Artificial VIEW *of the* TRANSIT *of* VENUS *June 3. 1769* at LONDON, *in a large* CELESTIAL LANDSCAPE.

A depiction of Benjamin Martin's 'Artificial Transit' – during the performances in his London shop, a hidden mechanism moved the sun across the sky and Venus across the sun

* Joseph Banks and his botanist Daniel Solander visited the observatory while the *Endeavour* was repaired in Jakarta and were so impressed that on their return to London, a neighbour remarked, 'they don't stop talking about the magnificent and well equipped observatory'.

Equipped with the best instruments, astronomers had concentrated to time Venus's entry and exit, but many other people across the world just wanted to see the rare celestial spectacle. In London more than fifty curious spectators squeezed into instrument-maker Benjamin Martin's shop to see the image of the transit projected on to the wall. In case of clouds, Martin provided other entertainment for his audience with an '*Artificial Transit*' – a seven-and-a-half by five-foot depiction of the sky over London on the day of the transit including a clock-type mechanism that moved a model of Venus across a painted sun.

Satirical Print: 'Viewing the Transit of Venus'

In the American colonies, encouraged by extensive reporting in local newspapers, hundreds of curious spectators also joined the event. In Providence, Rhode Island, the transit was observed by 'most of the Inhabitants', while the amateurs who assembled in Charleston 'were totally deprived' because of clouds. Wherever astronomers set up their telescopes, crowds assembled to see the tiny black dot through smoked glasses. Though astronomers remained glued to their telescopes even when clouds and rain prevented them from seeing anything, their audiences quickly ran out of patience. Some looked for better entertainment. When a thunderstorm obscured the sky in Leiden, in the Netherlands, one spectator resolved to see an 'earthly Venus' in the opera house instead. The observations there, he wrote to a friend, turned out to be successful and the singer 'looked like she would have allowed some immersion'. These earthly pleasures were also enjoyed by some 'young Bloods' in London who, a newspaper reported, after they had seen the black dot on the sun 'made a *Transit* into Covent Garden among a number of the said beautiful Planets' – an area of the city at that time well known for its prostitutes.

All those who had watched knew this would be the last time that Venus would cross the sun during their lifetime. When the transit was over, it would take another 105 years before the planet would return as a black dot. Now, once again, the monumental task of collecting and sharing the data would begin in the attempt to consolidate the different figures into the one number they were all looking for: the exact distance between the sun and the earth.

15

After the Transit

On 11 June 1769, eight days after his successful viewing of the transit, Chappe d'Auteroche fell ill with typhus. As the mission of San José del Cabo had turned into 'a scene of horror', Chappe continued to work like a man possessed. During the nights he observed the stars and planets, and during the days he nursed his team. Everybody was affected – the Spanish observers and their servants had managed to watch Venus but had then collapsed, as had Chappe's engineer, watchmaker and painter. Almost everyone in the mission was 'either dying or hastening towards death'. By the time Chappe himself woke with a high fever, there was no one healthy enough to help him. Trembling, he opened his trunk and tried to medicate himself with purgatives. He couldn't die yet, he resolved, because seven days later, on 18 June, he needed to observe a lunar eclipse in order to establish the longitude of San José – without which his impressive transit observations would be utterly useless.

Delirious with fever and shaken by intense headaches, Chappe stared through his telescope to take the measurements. Clinging to life, he forced himself night after night to his instruments, observing the stars and satellite eclipses. Eventually he finished his notes, and, when he knew that death was inevitable, placed his transit records in a small box. He was determined to provide the scientific world with his invaluable measurements even after death.

By then the village and mission had become a 'mere desert' – a ghost town, emptied by disease and death. The heat had become unbearable and the surviving inhabitants were incessantly attacked by insects. Of the few who recovered, most were struck down again by a second deadly attack.* Only two members of the French expedition team, the engineer and the painter, were able to fight off the disease. Chappe, who had always regarded life through rose-tinted lenses of hyperbole and excitement, was strangely calm – he was content and quietly happy with his achievements. As death neared he was like a 'true philosopher', the engineer said. On 1 August, with all the necessary astronomical observations complete, Chappe died peacefully.

He had been one of the few astronomers to see Venus's march across the sun twice, and the only one who had observed both transits successfully from beginning to end. He had achieved more than any of the other observers, but his task was not yet fully accomplished: it lay in the weakened hands of his remaining team to ensure that the scientific world received these unique observations. The engineer and the painter, exhausted and barely alive themselves, took possession of the only copy of Chappe's notes. As they buried the astronomer in the hardened soil of San José, they knew it was now up to them to deliver the precious data to the Académie, and to ensure that Chappe had not died in vain.

In Pondicherry, Le Gentil was coming to terms with his disastrous failure. He felt betrayed and dejected. For two long weeks after the transit, he could do nothing. When he picked up his pen to write to the Académie in Paris to report his misfortune, it fell out of his hand. His mood didn't improve when he heard that the sky which had been so 'cruel' to him in Pondicherry had been perfectly clear in Manila where he had originally planned to view the transit. Once again he felt cursed. No other astronomer had spent so many years chasing Venus. Now he would have to return to Paris empty-handed.

As Le Gentil planned his sorrowful journey home, the members of the learned societies of Europe were beginning to collect the data from the second transit. Once again scientists in Paris, London, Stockholm and St Petersburg zealously took up the enormous task of collating the results. Astronomers

---

* All the Spaniards died except Vicente de Doz who returned to Spain in summer 1770 to deliver his own observations to his king.

rushed to write up their results, and by the end of 1769 huge numbers of observations had been exchanged between the scientific societies. This time – encouraged by Catherine the Great's interest in the transit – the Russians proved to be the most efficient. Only three months after the transit fifty-one printed reports were dispatched from the Imperial Academy to colleagues and societies across Europe.

But none of the Russian observations had been completely satisfactory. In St Petersburg, the German astronomer Christian Mayer, who had only arrived two weeks before the transit, noted that his entry times were uncertain because Venus appeared distorted. The German astronomer in Orenburg reported a similar phenomenon, writing about the 'wave-like' edge of the planet. One of the Swiss astronomers on the Kola peninsula managed to take the entry times but clouds prevented him from seeing Venus's exit, while the other Swiss (who was observing a bit further west) saw only a curtain of permanent rain. Rumovsky's assistant, who had been ordered at the last minute to observe the transit on Kildin, had died. The only one of Catherine's astronomers who had succeeded in taking both entry and exit times was Rumovsky himself in Kola – but he had some doubts, as clouds had often disturbed his view.

As more and more reports arrived in St Petersburg from across Europe, it became apparent that the weather had been worse than during the first transit. An astronomer in Göttingen had been troubled by clouds, while the observers in four Danish cities had seen absolutely 'nothing' because of bad weather. The observations in Uppsala and Stockholm had also been disappointing. The transit day had started, Wargentin told his international colleagues, as 'one of the most beautiful days of the summer, but towards the evening, just as they were expecting Venus, clouds had drifted over the sun. Some of them had noted the entry times, but with the evening sun partly covered and low on the horizon, these varied by several seconds. The Swedes had stayed up all night in the hope of catching another glimpse of Venus in the early morning hours, but the sky had remained stubbornly overcast.

The French were also disappointed. The viewing at Louis XV's Château de la Muette near Paris was interrupted by a sudden shower – prompting the large crowd of sightseers to push into the observing pavilion with much 'noise and confusion'. In Paris Lalande failed to time the first contact because 'I was precisely in the place where the clouds came on twenty-five seconds too soon'.

Astronomers in Britain had also struggled. Nevil Maskelyne examined the British observations and found the results worse than he had hoped for. The seven astronomers who had timed Venus's entry time at the Royal Observatory in Greenwich were a staggering fifty-three seconds apart – a bad result in any case but even more so for Maskelyne, who was famed for observations that achieved 'a degree of correctness seldom equalled & never surpassed'. The differences were 'greater than I expected', he noted dryly. In London some observers had problems because plumes of smoke billowed from the thousands of chimneys in the metropolis – something the astronomers in Glasgow tried to counter by placing an announcement in the local newspaper 'begging the inhabitants . . . to put out their fires'.

In addition to the bad weather, it also quickly became obvious that most astronomers had encountered the same phenomena as they had during the first transit. The observations were peppered with remarks such as: the planet's edge 'bubbled', or shifted 'like the waves of a tempestuous sea'. Venus was described alternately as 'an apple connected by its stalk' to the edge of the sun, 'the neck of a Florence flask' and a 'pointed truffle'. Others had remarked that the planet seemed to be attached to the edge of the sun by a 'small shadow' or a 'dark thread' during the moments of entering and exiting. 'Venus's circular figure was disturbed', Maskelyne reported from Greenwich, while others described how the planet was 'misshapen' or 'ill defined'. Some had once again seen the luminous ring and almost all European observers complained about 'tremulous vapours'. Despite using the best instruments, observers in Stockholm, Uppsala, Paris, Greenwich, St Petersburg, Orenburg and elsewhere in Europe and Russia had all been hampered again by Venus's caprices: the so-called black drop effect, vapours and undulations.

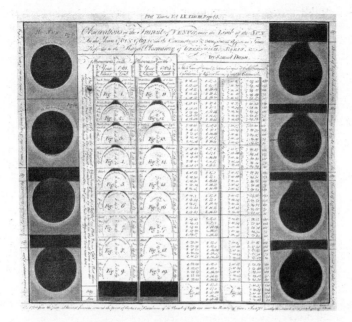

Drawings of the luminous ring around Venus and the black drop effect as
seen on 3 June 1769 in London

The results from Pello were no better, Wargentin discovered
when he received Fredrik Mallet's report in July. Attracting
misfortune almost as often as Le Gentil, poor Mallet had not
seen a thing. Though there had been clear skies for several
days previously, only hours before the transit clouds had
appeared. Mallet had missed the moment when Venus pushed
on to the sun in the evening as well as her exit in the early
morning hours. The melancholic astronomer could not believe
that he had travelled through ice and snow for absolutely
nothing. Every time he thought of the 'miserable night', Mallet
said, it made him even more depressed. He had 'fallen out'
with Venus, he told a friend.

Only the diligent Anders Planman had managed to observe
the transit in Kajana – despite the fact that the sky had been
hidden by a thick layer of clouds in the hours leading up to
it. The astronomer had been desperate. 'Never', he wrote to
Wargentin, 'have my fears and worries been so great'. Planman

had already given up all hope, when he saw with 'tears' in his eyes that the sky suddenly opened – just long enough for him to catch a glimpse of Venus moving on to the sun. With perfect timing the clouds closed in again only minutes after Planman had taken the external and internal entry times. Shortly afterwards he heard a thunderstorm rolling in, and presumed that he would never see Venus again. He had abandoned his observatory but fortunately returned at 2 a.m. because, as if pushed aside by a heavenly hand, the thick curtain of clouds opened again and the sun peeped through the grey sky – just at the moment when Venus exited. So far, Planman's observation was the only good one from the northern viewing stations.

Though no one had expected any great revelations from the observations in central Europe where most of the transit had happened in the hours of darkness, the early reports were disappointing. Now the scientists could only wait for the results from India, California, the South Sea, and the remaining reports from the Arctic Circle. Since almost every astronomer in Europe had struggled to take accurate data, expectations of the reports from the far-flung expeditions were low. It was very likely that astronomers on the other side of the globe had encountered the same problems, if not worse.

Despite the inaccuracies of the early results, what was truly impressive was the quick turnaround and efficient communication between the scientific communities. The Académie in Paris heard about the Stockholm and London observations at the end of June, while Planman's observations from Kajana arrived in early August in France. Wargentin's letter about the observations in Uppsala and Stockholm was read in the St Petersburg Academy less than six weeks after the transit. Taking into account ocean crossings and long overland travel times, it was extraordinary that Maskelyne was able to thank the North American colonists for their observations in Pennsylvania on 2 August, only two months after the transit and that Planman was reading what his Russian colleagues had experienced in September. In October a German newspaper printed the Russian observations and in November the first results from North America were published in the journal of the Royal Society. By the end of the year, the fellows of the Royal Society were informed by Lalande that both a French missionary in Martinique and Pingré in Haiti had succeeded in observing the beginning of the transit.*

In November 1769, more reports arrived from the north. Jeremiah Dixon in Hammerfest and William Bayley at the North Cape had both struggled with haze and clouds. William Wales, who had left Hudson Bay as quickly as possible after the transit, had been more successful. The many months of freezing weather had been worth it, he was able to tell the fellows of the Royal Society, because he had been able to record both the entry and exit times. Keen for the results from Vardø, the most important northern location, Lalande requested information from Maximilian Hell before the Jesuit had even returned home, but heard nothing back. Angered by Hell's apparent refusal, Lalande believed that the astronomer had failed to view the transit – a rumour that continued to circulate, fired by the increasing anti-Jesuit sentiment in Europe. The real reason for his silence was that Hell's first obligation was to the king of Denmark who had paid for the expedition. Christian VII insisted that Hell should present the results first to the Academy of Sciences in Copenhagen – which he did on 24 November 1769. Only then was the data from the Vardø observation allowed to be published and circulated, in a print run of 120 copies.†

The scientists of the learned societies had to wait an entire year to receive the next results. In December 1770, Chappe's engineer finally arrived in Paris with the notes of the successful California observations, and the news of the astronomer's death. Now only the times from Pondicherry and the South Sea were missing – but unbeknown to the European scientists,

* Pingré had only just made it to Haiti where he arrived on 23 May 1769, ten days before the transit.

† After his death, Hell was accused of having falsified his data by astronomer Carl Ludwig Littrow, who claimed that Hell had purposely delayed the publication. The proof of this, Littrow insisted in 1835, was that Venus's entry and exit times in Hell's manuscript were written in different ink. It turned out that Littrow was colour-blind and that he had been wrong about Hell. The observations in Vardø were some of the best of all the expeditions.

Le Gentil had failed to see anything, and the *Endeavour* crew, by then, was fighting death in Jakarta.*

Cook had left Tahiti a few weeks after the transit to embark on the second part of the mission. They had circumnavigated New Zealand and sailed along the east coast of Australia where Cook called the place of their first landfall Botany Bay – in tribute to the many new plants they discovered there. The botanists found towering eucalyptus trees with peeling bark and strange shrubs that paraded blossoms which looked like huge cones made of furry petals. The forests echoed with eerie noises from animals and birds no white man had ever seen or heard – a land that was promisingly abundant and fertile. As Joseph Banks and his botanist frantically collected plants (so many that they couldn't press them quickly enough), Cook made a survey of the coastline with Charles Green providing the astronomical observations to determine their exact positions.

Repairs to the *Endeavour* after the disastrous accident in the Great Barrier Reef

In the summer, as Chappe's engineer had travelled back to Europe, the *Endeavour* sailed into the Great Barrier Reef – 1,500 miles of labyrinthine reefs and coral islands – the most

* The only information the Royal Society had about the whereabouts of the *Endeavour* was two years old – from the letters which Cook had sent from Rio de Janeiro in December 1768.

treacherous stretch of coastline in the world. After only a few days the *Endeavour* ran aground. 'Fear of Death now stard us in the face', Banks wrote in his journal. With a huge hole in the hull, they barely made it to the shore, but after emergency repairs to the damaged ship, a determined Cook set out again to manoeuvre through the underwater maze of razor-sharp spurs, sandbanks and volatile currents.

Throughout these dangerous weeks of sailing, astronomer Charles Green remained steadfast at his instruments, even when the *Endeavour* was pushed on to walls of coral 'rising all most perpendicular out of the unfathomable Ocean'. Green calmly continued his observations, no matter if the *Endeavour* was about to be crushed 'to peices [sic]' as Cook feared. They eventually emerged battered but intact, and arrived in Jakarta two months later – the hull of the *Endeavour* ground down by the reefs and tropical worms to one-eighth of an inch. But just when they thought they had escaped, death began to stalk the crew: with its stagnant canals, Jakarta was infested with malaria and at that time the deadliest port in the world. When they left two-and-a-half months later, at the end of December 1770, Cook had lost more men than during the entire voyage before.

After an earthquake many of the canals in Jakarta became stagnant, turning them into a breeding ground for disease

The astronomers too would not all escape the terrible plague. On 29 January 1771, Charles Green died on board the *Endeavour* and was buried in the Indian Ocean. After the Russian observer in Kildin, Chappe and the Spanish observer Salvador de Medina in California, Green was the fourth transit astronomer to die on the quest to discover the size of the solar system. Green had been ill since their stay in Jakarta, but instead of resting had become so delirious that he made matters worse. 'In a fit of phrensy', newspapers later reported, 'he got up in the night and put his legs out of the portholes, which was the occasion of his death.' Even Cook, who always protected Green, admitted that the astronomer had brought his death upon himself because 'he took no care to repair' but on the contrary 'greatly promoted the disorders'.

Five-and-a-half months later, in July 1771, almost exactly three years after their departure, the *Endeavour* arrived back in Britain with the notes of the transit observations and trunks filled with pressed plants and drawings of exotic landscapes. They told stories of their astronomical observations and their amorous adventures in Tahiti, the dangers of their voyage and of the fertile land in Australia.

As Cook received a hero's welcome and astronomers pored over the transit notes, Le Gentil was still on his way to France. After the transit in Pondicherry, he had received a letter from his legal advisor in Normandy, informing him that his heirs were spreading rumours of his death so that they could divide up his estate. Le Gentil tried to leave immediately but before he could return home, he first had to sail to Mauritius where he had left his extensive natural history collections for safe-keeping. After months of waiting to find a passage and several almost fatal attacks of dysentery, he set out from Pondicherry on 1 March 1770. He disembarked in Mauritius six weeks later – a few days before Cook first sighted Australia – but was weakened from his long illness. It would be another seven months before Le Gentil felt strong enough to continue. When he did, he was immediately beset again: this time by a hurricane that forced the vessel back to Mauritius. By then the Frenchman had lost all belief that he would ever see France again. The sight of Mauritius, he cried, 'had become unbearable'.

On 30 March 1771, almost two years after the transit and

still in Mauritius, Le Gentil boarded a ship bound for Europe. He carried his instruments, notes and eight boxes full of natural history objects on board. He was desperate to return home but, by the time they rounded the Cape of Good Hope, Le Gentil was certain that his journey had come to an end. The sea was more wild, he said 'than I had ever seen before' – quite something from a man who had endured a never-ending chain of violent storms. With no energy or hope left, the disheartened Le Gentil spread his coat between the cargo trunks deep in the hold of the boat, lay down, closed his eyes and waited to die.

To his surprise, the ship survived, and on 1 August he sighted Cadiz in the distance. He decided now to take the overland route, because, he finally admitted, 'I was so tired of the sea'. Le Gentil was the last of the transit astronomers to set foot back on European soil. More than eleven years after he had left in search of Venus, he returned to Paris to find that his heirs had declared him dead and the Académie had removed him from their payroll.

Epilogue
A New Dawn

Once James Cook had safely returned to Britain with the transit results from Tahiti, astronomers had the all-important counterpart to the northern observations. Only the pernickety Nevil Maskelyne found something to complain about. Looking at Charles Green's observations, Maskelyne wrote that they 'differ more from one another than they ought to do'. Cook defended Green and remarked that the astronomer had died before he could edit his notes. The papers which had been presented to the Royal Society were Green's rough drafts, Cook said, asking if Maskelyne would 'publish to the world all the observations he makes good or bad or did he never make a bad observation in his life'.* Others were more pleased and excited to be finally able to calculate the distance between the earth and the sun. Though so many observations around the globe had failed, the ones from the South Sea were good enough to be useful.

Once again the astronomers of the learned societies in Britain, France, Sweden, Russia and America returned to their calculations – everybody wanting to be the first to present an accurate figure for the solar parallax. Lalande and Pingré were calculating in Paris, Maximilian Hell in Vienna, Anders Planman in Åbo, the members of the APS in Philadelphia, and the Swede Anders Lexell in Russia on behalf of the Imperial Academy of Sciences in St Petersburg – all rushing to publish their results in the scientific journals.

Five months after the return of the *Endeavour*, in December 1771, the fellows of the Royal Society met to hear the British

* Cook also pointed out that the observations had been difficult because the quadrant had been stolen by the Tahitians and taken apart just before the transit.

results. Taking the times from Maximilian Hell in Vardø, Stepan Rumovsky in Kola, William Wales in Hudson Bay, Chappe d'Auteroche in California as well as the Tahitian observations, Thomas Hornsby calculated that the solar parallax was 8"78, a figure very close to today's known value of 8"79.* The distance between the earth and the sun, Hornsby computed, was 93,726,900 miles – less than 800,000 miles off today's calculation of 92,960,000 miles. The 'uncertainty' that had riddled the 1761 observations was 'entirely removed', the British claimed. The learned societies 'may congratulate themselves' on succeeding in measuring the size of the solar system – or at least 'as accurate a determination', they admitted, 'as perhaps the nature of the subject will admit'.

The British optimism was a little exaggerated. Once again astronomers in different countries came to various parallax values. By including or excluding certain times and data, the calculations ranged from 8"43 to 8"80. Though not as precise as the astronomers had hoped, it was nevertheless a great improvement on the 1761 computations. Over the past two centuries astronomers had, bit by bit, whittled down the parallax. Kepler had estimated that it was less than 59" (which translated into a distance of less than 14,000,000 miles), Halley had predicted that it was no greater than 12"½, and the 1761 results had placed it between 8"28 to 10"60 (77,100,000 to 98,700,000 miles). An exact distance was still not agreed upon, but after 1769 the margins had substantially narrowed. Whereas the variations of the solar parallax in 1761 had amounted to more than 20,000,000 miles, astronomers were now able to reduce the difference to only 4,000,000 miles. After the 1769 transit, they had a much more precise idea of the real distance between the earth and the sun.

The achievements of the transit projects changed the world of science. Not only did they improve mankind's understanding of the solar system, but the by-products and benefits of the expeditions were numerous and wide-ranging. Maps, for

* Thomas Hornsby calculated the solar parallax by choosing a rather arbitrary selection of times – for example, he ignored Green's ingress time and took the mean of Cook's and Green's egress times – but his results were impressively close to today's value.

	Wardhus.		California.	
	Ingres. H. ′ ″	Egres. H. ′ ″	Ingres. H. ′ ″	Egres. H. ′ ″
Obferved times.	9 34 10,6	15 27 24,6	0 17 27,9	5 54 50,3
Effect of parallax.	+ 6 35,6	− 4 35,9	+ 24,9	+ 4 52
Reduced times.	9 40 46,2	15 22 48,7	0 17 52,8	5 59 42,3

	Kola.		Hudfon's Bay.	
	Ingreis.	Egrefs.	Ingrefs.	Egrefs.
Obferved times.	9 42 4	15 35 23	1 15 23,3	7 0 47
Effect of parallax.	+ 6 37,4	− 4 45,1	+ 4 15,9	+ 0 38,7
Reduced times.	9 48 41,4	15 30 37,9	1 19 39,2	7 1 25,7

King George's Ifland.

	Ingrefs.				Egrefs.
	Capt. Cook.	Mean.	Dr. Solander.	Mr. Green.	Mean.
Obferved times.	21 44 15,5	21 44 4,5	21 44 2,5	21 43 55,5	3 14 8
Effect of parallax.	5 40,4	5 40,4	5 40,4	5 40,4	+ 6 23,8
Reduced times.	21 38 35,1	21 38 24,1	21 38 22,1	21 38 15,1	3 20 31,8
Ditto at Wardhus.	9 40 46,2	9 40 46,2	9 40 46,2	9 40 46,2	15 22 48,7
Difference of meridians.	12 2 11,1	12 2 22,1	12 2 24,1	12 2 31,1	12 2 16,9

	Ingrefs.				Egrefs.
	Capt. Cook.	Mean.	Dr. Solander.	Mr. Green.	Mean.
Reduced times at K. G. Ifl.	21 38 35,1	21 38 24,1	21 38 22,1	21 38 15,1	3 20 31,8
Ditto at California.	0 17 52,8	0 17 52,8	0 17 52,8	0 17 52,8	5 59 42,3
Difference of meridians.	2 39 17,7	2 39 28,7	2 39 30,7	2 39 37,7	2 39 10,5

	Ingrefs.				Egrefs.
	Capt. Cook.	Mean.	Dr. Solander.	Mr. Green.	Mean.
Reduced times at K. G. Ifl.	21 38 35,1	21 38 24,1	21 38 22,1	21 38 15,1	3 20 31,8
Ditto at Kola.	9 48 41,4	9 48 41,4	9 48 41,4	9 48 41,4	15 30 37,9
Difference of meridians.	12 10 6,3	12 10 17,3	12 10 19,3	10 10 26,3	12 10 6,1

	Ingrefs.				Egrefs.
	Capt. Cook.	Mean.	Dr. Solander.	Mr. Green.	Mean.
Reduced times at K. G. Ifl.	21 38 35,1	21 38 24,1	21 38 22,1	21 38 15,1	3 20 31,8
Ditto at Hudfon's Bay.	1 19 39,2	1 19 32,9	1 19 39,2	1 19 39,2	7 1 25,7
Difference of meridians.	3 41 4,1	3 41 15,1	3 41 17,1	3 41 24,1	3 40 53,9

VOL. LXI.   4 E   The

A page from the report that was read to the fellows of the Royal Society, presenting the final results – this is a list of the times on which the calculations for the solar parallax were based example, had become more precise – from Le Gentil's new sea charts for the region around Madagascar and Fredrik Mallet's survey of the Gulf of Bothnia to new cartography in Russia.* Chappe had been the first to publish a book that included more than just his transit observations – *Voyage en Sibérie*. Le Gentil – once he had dealt with his wayward heirs and fought his way back into the employment of the Académie – wrote *Voyage dans les mers de l'Inde (1760–1771)* in which he described his odyssey and reported on climate, diseases, marine life, customs, wind patterns in the Indian Ocean and Brahmin astronomy. Jesuit Maximilian Hell also had plans for a three-volume publication which besides astronomy, he announced, would cover a wide range of subjects, including meteorology, the Northern Lights and reindeer herding. Hell's own observations would be complemented by information on northern flora, which was to be provided by the botanist who had accompanied him. At the same time Hell also planned to include his assistant János Sajnovics's revelatory discovery of the close relationship between the Hungarian and Sami languages (which to this day is regarded as the foundation of the study of the Uralic languages).†

Catherine the Great had also been ambitious. Together with the astronomers, she had dispatched teams of naturalists as well as taxonomists, painters and hunters, turning the astronomical expeditions into more wide-ranging journeys of scientific discovery. The German naturalist Peter Simon Pallas, for example, had been part of the Orenburg expedition on the Ural river for two explicit reasons: for the 'advantage of the empire' and the 'improvement of the sciences'. After the transit observation he continued to criss-cross Russia for five more years – during which time he also met several of the other transit astronomers including the German Georg Moritz Lowitz, who was now busy conducting surveys of possible

* In spring 1770 the Academy in St Petersburg commissioned a new map of the empire based on the longitudinal calculations that had been carried out during the transit expeditions.
† Hell did work on the book but failed to publish it, probably because it was too ambitious but also because Pope Clement XIV suppressed the Society of Jesus in 1773, which left Hell with no Jesuit staff to assist him. Sajnovics's book was called *Demonstratio idioma Ungarorum et Lapponum idem esse* (1770).

canals to open up central Russia for trade. Lowitz was to be the fifth astronomer – after the Russian who went to Kildin, Chappe, de Medina, and Green – to die in the name of science. In the summer of 1774, the Cossack insurrection prompted the Academy in St Petersburg to call back all their scientific teams, but Lowitz vowed to remain and finish his work. He was captured, brutally tortured and killed in August that year.

When Pallas arrived back in St Petersburg in July 1774 (after wisely following the Academy's advice), he brought with him an enormous natural history collection, ethnographical reports and a wealth of information on agriculture, manufacturing, mines, ores, salt-works and forestry. Vast swathes of Russia that had never been scientifically studied before were revealed, presenting new economic opportunities, as well as detailed information on their local populations and indigenous plants and animals. Combined with the transit observations, the expedition changed the perception of Russia in Europe and still counts as the most important exploration of the country to this day. Catherine had achieved what she had set out to do: fulfilling her country's obligations in the transit observations, and creating the foundation of comprehensive scientific research in her country.

The *Endeavour* returned to Europe with 30,000 dried plant specimens – some 3,600 species – of which a staggering 1,400 were new to Britain's botanists. It was a testimony to the economic promises that these distant countries held. Nature had become a 'vast book of information' that could be used to a nation's advantage, Joseph Banks said. Inspired by the *Endeavour* voyage and the knowledge he had gained of the climate, flora and soil of Australia, Banks consequently became the greatest promoter of the colonisation of the country. He also personally selected useful seeds and gave agricultural advice when the First Fleet sailed to Botany Bay in 1787.

Banks became the president of the Royal Society in 1778 (a post he held for four decades), advised the government on colonial enterprises and turned Britain into a centre for the scientific study and economic exploitation of the world's flora. Having experienced global cooperation between scientific communities during the transit years, he became its greatest advocate. Even when France declared war on Britain in 1793, he continued to help French scientists whenever he could, providing passports, sharing specimens or making his vast library available to them. 'The science of two Nations may be at Peace', he said, 'while their Politics are at war' – and it was this peace of the sciences that proved to be of vital significance for the advancement of knowledge.

As the transit astronomers and their teams returned with trunks filled with pressed plants, seeds, minerals and stuffed animals as well as detailed reports on soil and surveys of geography, climate and customs, the idea of the modern scientific expedition was born. Banks might have claimed to be the first man with a scientific education to embark on a voyage of discovery when he set out on the *Endeavour*, but after the transits of Venus this type of expedition became the norm. From then on major explorations always included scientific teams or at least some members who had undergone scientific training – from Meriwether Lewis and William Clark, who received detailed scientific instruction before they embarked on the first ever overland journey across the whole of the North American continent in 1803, to Charles Darwin on the *Beagle* in the 1830s. Even Napoleon Bonaparte's army in Egypt was accompanied by a corps of almost 200 scholars including chemists, mathematicians, linguists and botanists.

The transit projects revealed the importance of international communication and collaboration. Never before had scientists and thinkers banded together on such a global scale – not even war, national interests or adverse conditions could stop them. The intensity of their commitment was unparalleled and the international ties it fostered remained in place long after the transits.

Even discoveries which countries could potentially have used against each other were now shared. When, in 1775, the Académie des Sciences in Paris and the French government sponsored a prize for a new process to obtain saltpetre, submission was open to foreign scientists – a surprising gesture given the fact that saltpetre was needed for the production of gunpowder. The Académie publicised the prize widely and sent information to their old transit colleagues at the Royal Society, the Swedish Academy of Sciences and the Imperial Academy in St Petersburg.

The interests of science had transcended national boundaries. Today we take this international cooperation for granted, yet tend to talk about such global projects as if they were purely a twentieth- and twenty-first-century phenomena, but the foundations of such organised collaborations were laid in the 1760s. For the first time governments had funded large-scale scientific projects, putting in place a model for future generations. The peaceful cooperation across so many countries, societies and individuals which had been involved in the transit observations proved how important exchange and collaboration was to the advancement of knowledge. The seeds of the global village in which we live today were sown in the decade of the transits, when intrepid astronomers across Europe came together to answer Edmond Halley's call.

On 5 and 6 June 2012 (depending on where you are), the tiny black circle of Venus will again traverse the face of the sun. This will be the last transit until December 2117, and we will therefore be the last people for more than a century to see the phenomenon that once inspired scientists from all over the world to work together. When we look up and see a planet almost as big as our own, dwarfed by perspective against the immensity of the sun, we do so standing on the shoulders of the hundreds of brave men who watched the exact same spectacle 250 years ago.

List of Observers 1761

Nationality/ Organising Nation	Observers (as named in records)	Location of Observation (historical and modern place names)
?	Braun	St Petersburg, Russia
American	John Winthrop and two assistants	St John's, Newfoundland, Canada
Austrian	Maximilian Hell, Joseph Edler von Herbert, Joseph Xavier Liesganig, Karl Scherffer, Antonio Steinkeller, Müller, César-François Cassini de Thury, Caroli Mastalier, Lysogorski, Ignaz Rain and Archduke Joseph	Vienna, Austria
Austrian	Ferenc Weiss	Tyrnau, Austria (modern Trnava, Slovakia)
Austrian	Felix Freiherr von Ehrmann zum Schlug	Wetzlas, Pölla, Austria
Britain	Anon.	Tranquebar (modern Tharangambadi), India
British	William Magee	Calcutta (modern Kolkata), India
British	Heberden	London, England
British	John Canton	London, England
British	Samuel Dunn	London, England
British	James Porter	Constantinople (modern Istanbul), Turkey
British	Mr Martin	Pondicherry (modern Puducherry), India
British	Mr Ferguson	Pondicherry (modern Puducherry), India
British	Robert Barker	Pondicherry (modern Puducherry), India
British	Alexander Simpson	Basque Roads, Bay of Biscay

British	Joseph Harris	Brecknockshire, Wales
British	Goodwin	Oxford, England
British	Webster	Huntingdonshire, England
British	John Rotheram	Newcastle, England
British	Dunthorn	England
British	Harding	Bombay (modern Mumbai), India
British	Benjamin Martin	London, England
British	Nevil Maskelyne and Robert Waddington	St Helena
British	William Chapple	Powderham Castle, England
British	Charles Mason and Jeremiah Dixon	Cape of Good Hope, South Africa
British	John Ellicot and John Dolland	London, England
British	Richard Haydon	Liskeard, Cornwall, England
British	Earl Ferrers	Stanton, Leicestershire, England
British	Bartholomew Plaisted	Islamabad, India (modern Pakistan)
British	William Hirst, George Pigott and Mr Call	Madras (modern Chennai), India
British	Nathaniel Bliss, Charles Green and John Bird	Greenwich, England
British	Thomas Hornsby, John Bartlett and Thomas Phelps	Shirburn Castle, Oxfordshire, England
British	Isaac Fletcher, George Bell and Elihu Robinson	Mosser, West Cumberland, England
British	James Short, Prince William, Prince Henry, Prince Frederick, John Blair, John Bevis, Duke of York and Lady Augusta	London, England
British	John Knott	Chittagong, India (modern Bangladesh)
Danish	Christian Horrebow	Copenhagen, Denmark
Danish	Bugge and Hascow	Trondheim, Norway
Dutch	Johannes Lulofs	Leiden, Netherlands
Dutch	Geradus Kuypers	Dordrecht, Netherlands
Dutch	Wytse Foppes Dongjuma	Camminghaburg, Leeuwarden, Netherlands
Dutch	Martinus Martens	Amsterdam, Netherlands
Dutch	Dirk Klinkenberg	Catshuis, The Hague, Netherlands
Dutch	Gerrit de Hahn, Pieter Jan Soele and Johan Maurits Mohr	Batavia, (modern Jakarta), Indonesia
Dutch	Jan de Munck	Middelburg, Netherlands

French	Alexandre-Gui Pingré and Denis Thuillier	Rodrigues
French	Duchoiselle	Grand Mount, near Madras (modern Chennai), India
French	de Seligny	Mauritius
French	Béraud	Lyon, France
French	Cardinal de Luynes	Sens, France
French	Jeaurat de Barros	Paris, France
French	Prolange	Vincennes, France
French	L'Abbé Outhier	Bayeux, France
French	Jean-Baptiste Chappe d'Auteroche and two assistants	Tobolsk, Russia
French	Pierre-Charles Le Monnier, King Louis XV and Charles Marie de la Condamine	Château de Saint-Hubert, France
French	De Manse, Clauzade, Jean Bouillet, Jean-Henri-Nicolas Bouillet, Joseph-Bruno de Bausset de Roquefort and l'Evêque	Béziers, France
French	Charles Messier, Libour and A. H. Baudouin	Paris, France
French	de Merville, Clouet	Paris, France
French	Jean Bouin and Jarnard Bouin & Vincent Dulague	Rouen, France
French	Barthelemy Tandon, Jean-Baptiste Romieu and Etienne-Hyacinthe de Rotte	Montpellier, France
French	L'Abbé Nicolas de La Caille, Jean-Sylvian Bailly and Turgot de Brucourt	Conflans-sous-Carrière, France
French	Forneu	La Meule, France
French	Gautier	Vire, France
French	Dange	Lorient, France
French	Dollier or Dollières	Peking (modern Beijing), China
French	Jérôme de Lalande	Paris, France
French	Guillaume Le Gentil	Indian Ocean, ship
French/Italian	Jean-Dominique Maraldi, Zannoni and Belléri	Paris, France
French/ Portuguese	Joseph-Nicolas Delisle and Jose Joaquim Soares de Barros e Vasconcelos	Paris, France
French/Swedish	Benedict Ferner, Jean-Paul Grandjean de Fouchy, Noël, Baër and Passement	Château de la Muette, France
German	Schöttl	Laibach, Germany (modern Llubljana, Slovenia)

German	Gottfried Heinsius	Leipzig, Germany
German	Georgio Kratz and two observers	Ingolstadt, Germany
German	R. P. Hauser	Dillingen, Germany
German	Tobias Mayer	Göttingen, Germany
German	Georg Friedrich Kordenbusch	Nuremberg, Germany
German	Anon.	Regensburg, Germany
German	Christian Rieger	Madrid, Spain
German	Lampert Hinrich Röhl	Greifswald, Germany
German	Franz Huberti and one other observer	Würzburg, Germany
German	Johann Georg Palitzsch	Prohlis near Dresden, Germany
German	Friedrich Wilhelm Eichholz	Halberstadt, Germany
German	Anon.	Wittenberg, Germany
German	Professor Haubold	Dresden, Germany
German	Buck	Königsberg, Germany (modern Kaliningrad, Russia)
German	Prosper Goldhofer	Polling, Germany
German	Jean Henri Samuel Formey	Berlin, Germany
German	Grafenhahn and Pöhlmann	Bayreuth, Germany
German	Christian Mayer and Elector Palatine Karl Theodor	Schwetzingen, Germany
German	Professor Polack and one other observer	Frankfurt an der Oder, Germany
German	Georg Christoph Silberschlag, Marktgraf Heinrich and Heinrich Wilhelm Bachmann	Kloster Berge, near Magdeburg, Germany
German	Georg Friedrich Brander, Peter von Osterwald, Johann Georg von Lori, Johann Georg Dominicus von Linbrunn and Ildephons Kennedy	Nymphenburg Palace, Munich, Germany
Germany	Anon.	Meissen, Germany
German	Eugen Dobler and Bertholdi	Kremsmünster, Austria
German	Itanow	Danzig, Germany (modern Gdansk, Poland)
Italian	Giovanni Battista Audiffredi	Rome, Italy
Italian	Leonardo Ximenes	Florence, Italy
Italian	Agostino Salluzzo	Rome, Italy
Italian	Niccolo Maria Carcani	Naples, Italy
Italian	Giuseppe Maria Asclepi	Rome, Italy
Italian	Giovanni Poleni	Padua, Italy
Italian	Giovanni Magrini	Imola, Italy
Italian	Daniel Avelloni	Venice, Italy
Italian	Jacopo Belgrado	Parma, Italy

Italian	Tommaso Narducci and Sacchetti	Lucca, Italy
Italian	Spagnius, François Jacquier and Thomas Le Seur	Rome, Italy
Italian	Giovanni Battista Beccaria, Canonica and Revelli	Turin, Italy
Italian	Eustachio Zanotti, Petronio Matteucci, Marini, Frisius, Casali and Sebastiano Canterzani	Bologna, Italy
Malta	Anon.	Valetta, Malta
Polish	Stefan Luskina	Warsaw, Poland
Portuguese	Teodoro de Almeida	Porto, Portugal
Portuguese	Miguel António Ciera	Lisbon, Portugal
Russian	Nikita Popov, Ochtenski and Tartarinov	Irkutsk, Russia
Russian	Mikhail Lomonosov	St Petersburg, Russia
Russian	Stepan Rumovsky and assistant	Selenginsk, Russia
Russian	Theodor Soimonoff	Siberia, Russia
Russian	Andrey D. Krasilnikov and Nikolay Kurganov	St Petersburg, Russia
Spanish	Antonius Eximenus	Madrid, Spain
Spanish	Benevent	Madrid, Spain
Spanish	de Ronas	Manila, Philippines
Swedish	Anders Wikström and one other observer	Kalmar, Sweden
Swedish	Nils Schenmark and Johan Henrik Burmestern	Lund, Sweden
Swedish	Anders Planman, Frosterus, Lagus and Planman's younger brother	Cajaneborg (modern Kajana or Kajaani), Finland
Swedish	Johan Justander and Wallenius	Åbo (Turku in Finnish), Finland
Swedish	Nils Gissler, Ström and one other observer	Hernosand, Sweden
Swedish	Bergström and Zegollström	Karlskrona, Sweden
Swedish	Anders Hellant, Lagerbohm and Häggmann	Torneå (Tornio in Finnish), Finland
Swedish	Brehmer, Landberg and Dehn	Landskrona, Sweden
Swedish	Mårten Strömer, Fredrik Mallet, Daniel Melander and Torbern Olof Bergman	Uppsala, Sweden
Swedish	Pehr Wilhelm Wargentin, Johan Carl Wilcke, Samuel Klingenstierna, Jacob Gadolin, Queen Louisa Ulrika, Crown Prince Gustav, Johan Gabriel von Seth, Pehr Lehnberg and Carl Lehnberg	Stockholm, Sweden

List of Observers 1769

Nationality/ Organising Nation	Observers (as named in records)	Location of Observation (historical and modern place names)
American	James Browne, Stephen Hopkins and Benjamin West	Providence, Rhode Island, America
American	John Page	America
American	David Rittenhouse, William Smith, John Lukens and John Sellers	Norriton, Pennsylvania, America
American	John Winthrop	Cambridge, Massachusetts, America
American	Owen Biddle, Joel Bailey and Richard Thomas	Cape Henlopen, Lewes, Delaware, America
American	William Poole	Wilmington, Delaware, America
American	William Alexander	Baskenridge, New Jersey, America
American	Samuel Williams and Tristram Dalton	Newbury, Massachusetts, America
American	John Ewing, Joseph Shippen, Hugh Williamson, Charles Thomson, Thomas Prior and James Pearson	Philadelphia, Pennsylvania, America
American	John Leeds	Talbot County, Maryland, America
American	Manasseh Cutler	Massachusetts, America
American	Ezra Stiles, William Vernon, Henry Marchant, Benjamin King, Henry Thurston, Punderson Austin, Christopher Townsend, William Ellery and Caleb Gardner	Newport, Rhode Island, America
Britain	Anon.	Mussoorie, Himalaya Mountains, India
British	Charles Mason	Cavan, Ireland
British	William Wales and James Dymond	Prince of Wales Fort (modern Churchill), Hudson Bay, Canada

British	Alexander Rose	Phesabad (modern Faizabad), India
British	William Bayley	North Cape, Norway
British	Nevil Maskelyne, William Hirst, John Horsley, Samuel Dunn, Peter Dollond, Edward Nairne and Malachy Hitchins	Greenwich, England
British	James Horsfall	London, England
British	John Canton	London, England
British	Alexander Aubert	London, England
British	Daniel Harris, James Townley and Dr Bostock	Windsor Castle, England
British	Lord and Lady Macclesfield, John Bartlett and Thomas Phelps	Shirburn Castle, Oxfordshire, England
British	Ludlam	Norton, near Leicester, England
British	Lucas	Oxford, England
British	Clare	Oxford, England
British	Sykes	Oxford, England
British	Shuckburgh	Oxford, England
British	Thomas Hornsby	Oxford, England
British	Cyril Jackson and John Horsley	Oxford, England
British	John Bevis and Joshua Kirby	Kew, England
British	James Lind, James Hoy and Lord Alemoor	Hawkhill, Scotland
British	Francis Wollaston	East Dereham, England
British	John Smeaton	Austrope Lodge, near Leeds, England
British	Brice	Kirknewton, Scotland
British	Alexander Wilson, Dr Williamson, Dr Reid, Dr Irvine and P. Wilson	Glasgow, Scotland
British	Robinson	Hinckley, England
British	G.G.	Leyburn, England
British	John Bradley	Lizard Point, Cornwall, England
British	Lieutenant Alexander Jardine and two other observers	Gibraltar, Spain/ England
British	Thomas Wright	Île aux Coudres, near Quebec, Canada
British	Gilbert White	Selborne, England
British	Jeremiah Dixon	Hammerfest, Norway
British	King George III, Queen Charlotte, Stephen C. T. Demainbray, Stephen Rigaud, Justin Vulliamy and Ben Vulliamy	Richmond, England
British	Mr Call	Madras (modern Chennai), India

British	East India Company	Sumatra, Indonesia
British	Captain Williams	Copenhagen, Denmark
British	Six, Ridoubt and one other observer	Canterbury, England
British	Lionel Charlton	Whitby, England
British	Musgrave	Plymouth, England
British	Captain Saunders	near Archangel, Russia, ship
British	William Richardson	twenty miles from St Petersburg, Russia
British	Benjamin Martin	London, England
British/French	Samuel Holland and Mr St Germain	Quebec, Canada
British/Russian	Williamson and Nikitin	Oxford, England
British/Swedish	James Cook, Charles Green, Daniel Solander, John Gore, Jonathan Monkhouse, William Monkhouse, Herman Spöring, Zachary Hicks, Charles Clerk, Richard Pickersgill and Patrick Saunders	King George Island or Otaheite (modern Tahiti, French Polynesia)
Denmark	Anon.	Copenhagen, Denmark
Denmark	Anon.	Friedrichsberg, Denmark (modern Germany)
Denmark	Anon.	Tromsdalen, Norway
Denmark	Peder Horrebow and Ole Nicolai Bützow	Dønnes, Norway
Danish/ Austrian	Maximilian Hell, János Sajnovics and Jens Finne Borchgrevink	Wardhus or Wardoe (modern Vardø), Norway
Danish/German	Christian Gottlieb Kratzenstein	Trondheim, Norway
Dutch	Johan Maurits Mohr	Batavia (modern Jakarta), Indonesia
French	Tourneau	Laon, France
French	Prince de Croï	Calais, France
French	Desilrabelle	Marseille, France
French	de Mantial	Nancy, France
French	Guillaume Le Gentil	Pondicherry (modern Puducherry), India
French	Jean-Baptiste Chappe d'Auteroche, Pauly and Dubois	San José del Cabo, Baja California, Mexico
French	Alexandre-Gui Pingré, Claret de Fleurieu, de la Fillière and Destourès	Cape François, Saint Domingue or Santo Domingo (modern Haiti)
French	Louis Cipolla	Peking (modern Beijing), China
French	Collas	Peking (modern Beijing), China
French	Charles Messier, Boudouin, Turgot de Brucourt and Zannoni	Paris, France

French	César-François Cassini de Thury, Duc de Chaulnes, Achille-Pierre Dionis de Séjour and Jean-Dominique Maraldi	Paris, France
French	Gabriel de Bory, Jean-Paul Grandjean de Fouchy, Jean Sylvain Bailly, Noël and L'Abbé Bourriot	Château de la Muette, France
French	Pierre-Charles Le Monnier and Joseph-Bernard de Chabert	Château de Saint-Hubert, France
French	de Saron	Saron, France
French	Jean-Baptiste d'Après de Mannevillette	Château de Kergars, France
French	Edmé-Sébastian Jeaurat	Paris, France
French	Antoine Darquir de Pellepoix	Toulouse, France
French	François Garipuy	Toulouse, France
French	Bouin and Vincent Dulague	Rouen, France
French	Diquemar	Le Havre, France
French	L'Abbé Faugère and de la Rogue	Bordeaux, France
French	Christophe	Martinique
French	Jérôme de Lalande and l'Abbé Marie	Paris, France
French	Fortin, Blondeau and Pierre Le Roy, de Verdun	Brest, France
French/British	Louis Degloss, J. Lang and H. Stoker	Dinapore (modern Danapur), India
French/British	Nathan Pigott, Pigott, Jr and Rochefort	Caen, France
German	Lampert Hinrich Röhl and Andreas Mayer	Greifswald, Germany
German	Wenceslaus Johann Gustav Karsten	Bützow, Germany
German	Elector Palatine Karl Theodor and Prince Franz Xaver of Saxony	Schwetzingen, Germany
German	Gotthelf Kästner, Ljungberg and Lichtenberg	Göttingen, Germany
German	John Godefrey Kochler	Leipzig, Germany
German	Johan Elert Bode	Hamburg, Germany
German	Jean Henri Samuel Formey	Berlin, Germany
German	Ackermann	Kiel, Germany
German/Russian	Christoph Euler and assistant	Orsk, Russia
German/Russian	Wolfgang Ludwig Krafft and assistant	Orenburg, Russia
German/Russian	Georg Moritz Lowitz and Pjotr Inochodcev	Guryev, Russia (modern Atyrau, Kazakhstan)
German/Russian	Catherine the Great, Franz Aepinus and eighteen courtiers	Oranienburg, Russia
German/Russian	Christian Mayer, Anders Johan	St Petersburg, Russia

	Lexell, Stahl, Johann Albrecht Euler and Andrey D. Krasilnikov	
German/ Swedish?	Brashe and two other observers	Lübeck, Germany
Italian	Sebastiano Canterzani	Bologna, Italy
Russian	Islenieff	Yakutsk, Russia
Russian	Stepan Rumovsky and Brolodin (or Borodulin)	Kola, Russia
Russian	Ochtenski	Kildin, Russia
Spanish	de Queiros	?
Spanish	Vicente Tolfino	Cadiz, Spain
Spanish	Vicente de Doz and Salvador de Medina	San José del Cabo, Baja California, Mexico
Spanish	Joaquin Velázquez de León	Santa Anna, Baja California, Mexico
Spanish	José Ignacio Bartolache, José Antonio Alzate and Antonio de Léon y Gama	Mexico City, Mexico
Spanish/Italian	Don Estevan y Melo and one Italian observer	Manila, Philippines
Swedish	Johan Törnsten	Frösön, Sweden
Swedish	Anders Planman and Uhlwyk	Cajaneborg (modern Kajana or Kajaani), Finland
Swedish	Johan Gadolin and Johan Justander	Vanhalinna, near Åbo (Turku in Finnish), Finland
Swedish	Pehr Wilhelm Wargentin, Benedict Ferner, Johan Carl Wilcke and Strussenfelt	Stockholm, Sweden
Swedish	Nils Schenmark and Olof Nenzelius	Lund, Sweden
Swedish	Fredrik Mallet	Pello, Finland
Swedish	Anders Hellant	Torneå (Tornio in Finnish), Finland
Swedish	Nils Gissler and Ström	Hernosand, Sweden
Swedish	Erik Prosperin, Daniel Melander, Salenius, Mårten Strömer and Torbern Olof Bergman	Uppsala, Sweden
Swedish	Johan Henrik Lidén	Leiden, Netherlands
Swiss	Johann Bernoulli	Colombes, France
Swiss/Russian	Jacques André Mallet	Ponoy, Russia
Swiss/Russian	Jean-Louis Pictet	Umba, Russia

Selected Bibliography,
Sources and Abbreviations

Abbreviations Archives and Sources

APS	American Philosophical Society, Philadelphia
Banks Journal	*The Endeavour Journal of Joseph Banks, 1768– 71,* online
BF online	Papers of Benjamin Franklin online
BL	British Library
CMRS	Council Meetings, Royal Society, London
Cook Journal	*James Cook's Journal of Remarkable Occurrences aboard His Majesty's Bark Endeavour, 1768–71,* online
DLC	Library of Congress, Washington DC
Histoire &	*Histoire de l'Académie Royale des Sciences . . .*
Mémoires	*Avec les Mémoires de Mathématique & de Physique,* l'Académie Royale des Sciences, Paris
JBRS	Journal Books, Royal Society, London
KVA	Kungliga Vetenskapsakademien, Stockholm (Swedish Royal Academy of Sciences)
KVA Abhandlungen	*Der Königl. Schwedischen Akademie der Wissenschaften Abhandlungen aus der Naturlehre, Haushaltskunst und Mechanik* (German edition of the KVA journal)
KVA Protocols	Protocols of the KVA meetings, Center for History of Science, Stockholm
Phil Trans	*Philosophical Transactions,* Royal Society, London
Protocols	Veselovsky, Konstantin Stepanovich (ed.), *Protokoly zasedaniy konferentsii Imperatorskoy Akademii nauk s 1725 po 1803 goda,* St Petersburg, Imperial Academy of Sciences, 1897–1911 (Protocols of the meetings of the Imperial Academy of Science, St Petersburg)
PV Académie	Procès-Verbaux online, Académie des Sciences, Paris
RGO	Royal Greenwich Observatory archives, Cambridge University Library, Cambridge

RS	Royal Society, London
RS L&P	Letters and Papers, Royal Society, London
RS MM	Miscellaneous Manuscripts, Royal Society, London

Online Sources and Internet Archives

Almost every article and report that was published at the time of the 1761 and 1769 transits is listed on Rob van Gent's online bibliography (often with a direct link to the original text).
http://transitofvenus.nl/wp/past-transits/bibliography-1761–1769/

The Endeavour Journal of Joseph Banks, 1768–71
http://southseas.nla.gov.au/index_voyaging.html

James Cook's Journal of Remarkable Occurrences aboard His Majesty's Bark Endeavour, 1768–71
http://southseas.nla.gov.au/index_voyaging.html

Histoire de l'Académie Royale des Sciences . . . Avec les Mémoires de Mathématique & de Physique, l'Académie Royale des Sciences, Paris
http://gallica.bnf.fr/ark:/12148/cb32786820s/date.r=.langEN

Procès-Verbaux, Académie des Sciences, Paris
http://gallica.bnf.fr/Search?adva=1&adv=1&tri=title_sort&t_relation=% 22Notice+d%27ensemble+%3A+http%3A%2F%2Fcatalogue.bnf. fr%2Fark%3A%2F 12148%2Fcb375720275%22&q=Procès-Verbaux+ +l%27Académie+Royale+des+Sciences&lang=en

Der Königl. Schwedischen Akademie der Wissenschaften Abhandlungen aus der Naturlehre, Haushaltskunst und Mechanik
http://gdz.sub.uni-goettingen.de/dms/loadoc/?PPN=PPN324352840

Philosophical Transactions, Royal Society, London
http://rstl.royalsocietypublishing.org/content/by/year

Göttingische Anzeigen von Gelehrten Sachen, Göttingen
http://gdz.sub.uni-goettingen.de/dms/load/toc/?IDDOC=66088

Das Neuste aus der Anmuthigen Gelehrsamkeit, Leipzig
http://gdz.sub.uni-goettingen.de/dms/load/toc/?PPN=PPN556861817

The Papers of Benjamin Franklin
http://franklinpapers.org/franklin/

Roode, van Steven, Transit Times (invaluable calculator that provides the exact transit times from 1639 to 2117)
http://transitofvenus.nl/wp/where-when/local-transit-times/

Newspapers and Journals

Quoted newspapers and journals are referenced in the endnotes. For accounts of the transit observations the following are useful:

Britain:
Philosophical Transactions, Royal Society, London

France:
Histoire de l'Académie Royale des Sciences . . . Avec les Mémoires de Mathématique & de Physique, l'Académie Royale des Sciences, Paris

Germany:
Göttingische Anzeigen von Gelehrten Sachen, Göttingen
Das Neuste aus der Anmuthigen Gelehrsamkeit, Leipzig

Sweden:
Kungl. Svenska Vetenskapsakademiens handlingar, Stockholm
Or the German edition: *Der Königl. Schwedischen Akademie der Wissenschaften Abhandlungen aus der Naturlehre, Haushaltskunst und Mechanik*, Stockholm

Russia:
Novi Commentarii, Imperial Academy of Sciences, St Petersburg

Italy:
Novelle Letterarie, Florence

Books

Acerbi, Joseph, Travels through Sweden, Finland, and Lapland, to the North Cape, in the years 1798 and 1799, London, printed for J. Mawman, 1802

Alder, Ken, *The Measure of All Things*, London, Little, Brown, 2002

Anon., 'Du passage de Vénus sur le Soleil, Annoncé pour l'année 1761', *Histoire & Mémoires*, 1757

Anon., 'Du passage de Vénus sur le Soleil, Qui s'observera en 1769', *Histoire & Mémoires*, 1757

Anon., 'Éloge de M. de l'Isle', *Histoire & Mémoires*, 1768

Anon., 'Sur la conjonction écliptique de Vénus & du Soleil, du 6 Juin 1761', *Histoire & Mémoires*, 1761

Anon., 'Sur la conjonction écliptique de Vénus et du Soleil, du 3 Juin 1769', *Histoire & Mémoires*, 1769

Anon., 'Sur le passage de Vénus sur le Soleil, du 3 Juin 1769', *Histoire & Mémoires*, 1770

Armitage, Angus, 'Chappe d'Auteroche. A Pathfinder for Astronomy', *Annals of Science*, 1954, vol. 10

——*Edmond Halley*, London, Nelson, 1966

——'The Pilgrimage of Pingré', *Annals of Science*, 1953, vol. 9

Aspaas, Per Pippin, 'Maximilian Hell's Invitation to Norway', *Communications in Asteroseismology*, 2008, vol. 149

Aubert, Alexander, 'Transit of Venus Over the Sun, Observed June 3, 1769', *Phil Trans*, 1769, vol. 59

Aughton, Peter, *Endeavour. Captain Cook's First Great Voyage*, London, Phoenix, 2003

Bacmeister, H. L. C., *Russische Bibliothek*, St Petersburg, Riga, Leipzig, Johann Friedrich Hartnoch, 1772, vol. 1

Bayley, William, 'Astronomical Observations Made at the North Cape, for the Royal Society', *Phil Trans*, 1769, vol. 59

Beaglehole, J. C. (ed.), *The Endeavour Journal of Joseph Banks, 1768–1771*, Sydney, Angus and Robertson, 1962

——*The Journals of Captain James Cook on his Voyages of Discovery*, Woodbridge, Suffolk, Boydell Press, 1999

Bernoulli, Jean, *Lettres astronomiques, ou l'on donne une idée de l'état actuel de l'astronomie pratique dans plusieurs villes de l'Europe*, Berlin, 1771

BOOKS

Bevis, John, 'Observations of the Last Transit of Venus, and of the Eclipse of the Sun the Next Day; Made at the House of Joshua Kirby, Esquire, at Kew', *Phil Trans*, 1769, vol. 59

Biddle, Owen, Joel Bailey and Richard Thomas, 'An Account of the Transit of Venus over the Sun, June 3d, 1769, as Observed near Cape Henlopen, on Delaware', *Transactions APS*, 1769–71, vol. 1

Biddle, Owen, 'Observations of the Transit of Venus Over the Sun, June 3, 1769; Made by Mr Owen Biddle and Mr Joel Bayley, at Lewestown, in Pennsylvania. Communicated by Benjamin Franklin', *Phil Trans*, 1769, vol. 59

Black, Jeremy, *George III. America's Last King*, New Haven and London, Yale University Press, 2006

Bliss, Nathaniel, 'Observations on the Transit of Venus over the Sun, on the 6th of June 1761', *Phil Trans*, 1761–62, vol. 59

Bode, Johann Elert, *Deutliche Abhandlung nebst einer Allgemeinen Charte von dem bevorstehenden merkwürdigen Durchgang der Venus durch die Sonnenscheibe*, Hamburg, 1769

Bradley, James, *Miscellaneous Works and Correspondence*, Oxford, Oxford University Press, 1832

Bradley, John, 'Some Account of the Transit of Venus and Eclipse of the Sun, as Observed at the Lizard Point, June 3d, 1769', *Transactions APS*, 1769–71, vol. 1

Brasch, Frederick, 'John Winthrop (1714–1779). America's First Astronomer, and the Science of his Period', *Publications of the Astronomical Society of the Pacific*, 1916, vol. 28

Candaux, Jean-Daniel (ed.), *Deux astronomes genevois dans la Russie de Catherine II: journaux de voyage en Laponie russe de Jean-Louis Pictet et Jacques-André Mallet pour observer le passage de Vénus devant le disque solaire, 1768–1769*, Ferney-Voltaire, Centre international d'étude du XVIIIe siècle, 2005

Canton, John, 'A Letter to the Astronomer Royal, from John Canton, M.A.F.R.S. Containing His Observations of the Transit of Venus, June 3, 1769, and of the Eclipse of the Sun the Next Morning', *Phil Trans*, 1769, vol. 59

——'Observations on the Transit of Venus, June the 6th, 1761, Made in Spital-Square', *Phil Trans*, 1761–62, vol. 52

Carlid, Göte and Johan Nordström (ed.), *Torbern Bergman's Foreign Correspondence*, Stockholm, Almqvist & Wiksell, 1965

Carter, Harold, *Sir Joseph Banks (1743–1820)*, London, British Museum, 1988

——'The Royal Society and the Voyage of HMS "Endeavour" 1768–71', *Notes and Records of the Royal Society of London*, 1995, vol. 49

Cassini, Jean-Dominique, *Éloge de M. Le Gentil*, Paris, de D. Colas, 1810

Catherine II, *The Antidote; or an Enquiry into the Merits of a Book, entitled A Journey into Siberia*, London, 1772

Chabert, Joseph-Bernard de, 'Mémoire sur l'avantage de la position de quelques isles de la mer de Sud, pour l'observation de l'entrée de Vénus devant le Soleil, qui doit arriver le 6 Juin 1761', *Histoire & Mémoires*, 1757a

——'Mémoire sur la nécessité, les avantages, les objets & les moyens d'exécution du voyage que l'Académie propose de faire entreprendre à M. Pingré dans la partie occidentale & méridionale de l'Afrique, à l'occasion du passage de Vénus devant le Soleil, qui arrivera le 6 Juin 1761', *Histoire & Mémoires*, 1757b

Chambers, Neil, *The Letters of Sir Joseph Banks. A Selection, 1768–1820*, London, Imperial College Press, 2000

Chaplin, Joyce E., *The Scientific American: Benjamin Franklin and the Pursuit of Genius*, New York, Basic Books, 2006

Chappe d'Auteroche, Jean-Baptiste, *A Journey into Siberia*, London, T. Jefferys, 1770

——*A Voyage to California to Observe the Transit of Venus*, London, Edward and Charles Dilly, 1778

——'Extract of a Letter, Dated Paris, Dec. 17, 1770, to Mr Magalhaens, from M. Bourriot; Containing a Short Account of the Late Abbé Chappe's Observation of the Transit of Venus, in California', *Phil Trans*, 1770, vol. 60

——*Mémoire du passage de Vénus sur le Soleil: Contenant aussi quelques autres Observations sur l'Astronomie, et la Déclinaison de la Boussole, faites à Tobolsk en Sibérie l'Année 1761*, St Petersburg, Imperial Academy of Sciences, 1762

——*Voyage en Sibérie, fait par ordre du roi en 1761*, Paris, Debure, 1768

Cipolla, Louis, 'Astronomical Observations by the Missionaries at Pekin', *Phil Trans*, 1774, vol. 64

Cohen, I. Bernard, *Science and the Founding Fathers: Science in the Political Thought of Thomas Jefferson*, Benjamin Franklin, John Adams, and James Madison, New York and London, W. W. Norton, 1995

Cook, Alan, *Edmond Halley: Charting the Heavens and the Seas*, Oxford, Clarendon Press, 1998

Cook, James, 'Observations Made, by Appointment of the Royal Society, at King George's Island in the South Sea; By Mr Charles Green, Formerly Assistant at the Royal Observatory at Greenwich, and Lieut. James Cook, of His Majesty's Ship the *Endeavour*', *Phil Trans*, 1771, vol. 61

Cope, Thomas D. and H. W. Robinson, 'Charles Mason, Jeremiah Dixon and the Royal Society', *Notes and Records of the Royal Society of London*, 1951, vol. 9

Cope, Thomas D., 'Some Contacts of Benjamin Franklin with Mason and Dixon and Their Work', *Proceedings APS*, 1951, vol. 95

——'The First Scientific Expedition of Charles Mason and Jeremiah Dixon', *Pennsylvania History*, 1945, vol. 7

Coxe, William, *Travels into Poland, Russia, Sweden, and Denmark*, London, T. Cadell, 1784

Croarken, Mary, 'Astronomical Labourers: Maskelyne's Assistants at the Royal Observatory, Greenwich, 1765–1811', *Notes and Records of the Royal Society of London*, 2003, vol. 57

Crosby, B., *Authentic Memoirs of the Life and Reign of Catherine II*, London, 1797

Cross, A. G. (ed.), *An English Lady at the Court of Catherine the Great: The Journal of Baroness Elizabeth Dimsdale, 1781*, Northampton, Crest Publications, 1989

Crump, Thomas, A *Brief History of Science, as Seen Through the Development of Scientific Instruments*, London, Constable & Robinson, 2001

Delambre, Jean-Baptiste-Joseph, 'Notice sur la vie et les ouvrages de M. Messier', *Histoire & Mémoires*, 1817

Delisle, Joseph-Nicolas, *La Description et l'usage de la mappemonde dressée pour le passage de Vénus sur le disque du Soleil qui est attendu le 6 juin 1761*, Paris, 1760

——*Mémoire présenté au Roi le 27 avril 1760 pour servir d'explication à la Mappemonde présentée en même temps à sa Majesté au sujet du passage de Vénus sur le Soleil, que l'on attend le 6 juin 1761*, Paris, 1760

Dixon, Jeremiah, 'Observations Made on the Island of Hammerfest, for the Royal Society', *Phil Trans*, 1769, vol. 59

Dixon, Simon, *Catherine the Great*, London, Profile, 2010

Donnert, Erich (ed.), *Europa in der Frühen Neuzeit. Festschrift für Günther Mühlpfordt*, Cologne, Böhlau Verlag, 2002

Dunn, Samuel, 'A Determination of the Exact Moments of Time When the Planet Venus Was at External and Internal Contact with the Sun's Limb, in the Transits of June 6th, 1761, and June 3d, 1769', *Phil Trans*, 1770, vol. 60

——'Some Observations of the Planet Venus, on the Disk of the Sun, June 6th, 1761', *Phil Trans*, 1761–62, vol. 52

Duyker, Edward and Per Tingbrand (eds.), *Daniel Solander. Collected Correspondence, 1753–1782*, Oslo, Copenhagen, Stockholm, Scandinavian University Press, 1995

'Early Proceedings of the American Philosophical Society for the Promotion of Useful Knowledge, Compiled by One of the Secretaries, from the Manuscript Minutes of Its Meetings from 1744–1838', *Proceedings APS*, 1885, vol. 22

East India Company, 'Observations Made at Dinapoor, June 4, 1769, on the Planet Venus, When Passing Over the Sun's Disk, June 4, 1769, with Three Different Quadrants, and a Two Foot Reflecting Telescope', *Phil Trans*, 1770, vol. 60

Encke, Johann Franz, *Der Venusdurchgang von 1769 als Fortsetzung der Abhandlung über die Entfernung der Sonne von der Erde*, Gotha, Beckerschen Buchhandlung, 1824

——*Die Entfernung der Sonne von der Erde aus dem Venusdurchgange von 1761*, Gotha, Beckerschen Buchhandlung, 1822

Engstrand, Iris H. W., *Spanish Scientists in the New World: The Eighteenth-century Expeditions*, Seattle, University of Washington Press, 1981

Euler, Christoph, *Auszug aus den Beobachtungen welche zu Orsk bey Gelegenheit des Durchgangs der Venus vorbey der Sonnenscheibe angestellt worden sind*, St Petersburg, Imperial Academy of Sciences, 1769

Ewing, John, 'An Account of the Transit of Venus over the Sun, June 3d, 1769, and of the Transit of Mercury Nov. 9th, Both as Observed in the State- House Square', Philadelphia, *Transactions APS*, 1769–71, vol. 1

Ferguson, James, 'A Delineation of the Transit of Venus Expected in the Year 1769', *Phil Trans*, 1763, vol. 53

——A *Plain Method of Determining the Parallax of Venus by her Transit over the Sun*, London, 1761

——*Astronomy Explained upon Sir Isaac Newton's Principles*, London, A. Millar 1764 (1st edition 1760)

Ferner, Benedict, 'An Account of the Observations on the Same Transit Made in and Near Paris', *Phil Trans*, 1761–62, vol. 52

——'Extract of a Letter to the Reverend Nevil Maskelyne, Astronomer Royal, from Mr Benedict Ferner, 9 June 1769', *Phil Trans*, 1769, vol. 59, p.404

Forbes, Eric Gray, 'Tobias Mayer (1723–62): A Case of Forgotten Genius', *The British Journal for the History of Science*, 1970, vol. 5

Fouchy, Jean-Paul Grandjean de, 'Éloge de M. L'Abbé Chappe', *Histoire & Mémoires*, 1769

Frängsmyr, Tore (ed.), *Science in Sweden: The Royal Swedish Academy of Sciences, 1739–1989*, Canton, Massachusetts, Science History Publications, 1989

Gadolin, Jacob, 'Beobachtungen beym Eintritte der Venus in die Sonne, den 3.Jun 1769 zu Åbo angestellt und eingegeben', *Der Königl. Schwedischen Akademie der Wissenschaften Abhandlungen aus der Naturlehre, Haushaltskunst und Mechanik*, 1769

Gascoigne, John, *Joseph Banks and the English Enlightenment: Useful Knowledge and Polite Culture*, Cambridge, Cambridge University Press, 1994

Gillispie, Charles Coulston (ed.), *Dictionary of Scientific Biography*, New York, Scribner, 1970–80

Gissler, Nils, 'Eintritt der Venus in die Sonne, den 3ten Jun. 1769 zu Hernosand', *Der Königl. Schwedischen Akademie der Wissenschaften Abhandlungen aus der Naturlehre, Haushaltskunst und Mechanik*, 1769

Gorbatov, Inna, *Catherine the Great and the French philosophers of the Enlightenment: Montesquieu, Voltaire, Rousseau, Diderot and Grim*, Bethesda, Academica Press, 2006

Greene, Jack P., *The Intellectual Construction of America: Exceptionalism and Identity from 1492 to 1800*, Chapel Hill and London, University of North Carolina Press, c. 1993

Hahn, Roger, *The Anatomy of a Scientific Institution: The Paris Academy of Sciences, 1666–1803*, Berkeley, University of California Press, 1971

Halley, Edmond, 'Halley's Dissertation on the method of finding the Sun's parallax and distance from Earth, by the Transit of Venus over the Sun's disk, June the 6th, 1761, originally published in Latin in 1716 in the Philosophical Transactions, translated to English', Ferguson, James, *Astronomy Explained upon Sir Isaac Newton's Principles*, London, A. Millarm 1764 (1st edition 1760)

——'Methodus Singularis Quâ Solis Parallaxis Sive Distantia à Terra, ope Veneris intra Solem Conspiciendoe, Tuto Determinari Poterit', *Phil Trans*, 1714–16, vol. 29

——'The Art of Living under Water: Or, a Discourse concerning the Means of Furnishing Air at the Bottom of the Sea, in Any Ordinary Depths', *Phil Trans*, 1714–16, vol. 29

Hamel, Jürgen, Isolde Müller and Thomas Posch, *Die Geschichte der Universitätssternwarte Wien. Dargestellt anhand ihrer historischen Instrumente und eines Manuskripts von Johann Steinmayr*, Frankfurt am Main, Harri Deutsch Verlag, 2010

Hansen, Truls Lynne and Per Pippin Aspaas, 'Maximilian Hell's Geomagnetic Observations of 1769 in Norway', *Tromsø Geophysical Observatory Reports*, no. 2, University of Tromsø, 2005

Harris, Daniel, 'Observations of the Transit of Venus Over the Sun, Made at the Round Tower in Windsor Castle, June 3, 1769', *Phil Trans*, 1769, vol. 59

Haydon, Richard, 'An Account of the Same Transit', *Phil Trans*, 1761–62, vol. 52

Hell, Maximilian, *Observatio Transitus Veneris Ante Discum Solis die 5ta Junii 1761*, Vienna, 1761

——*Observatio Transitus Veneris Ante Discum Solis die 3 junii anno 1769*, Giese, Copenhagen, 1770

Hellant, Anders, 'Venus in der Sonne zu Torne, den 6 Junii 1761', *Der Königl. Schwedischen Akademie der Wissenschaften Abhandlungen aus der Naturlehre, Haushaltskunst und Mechanik*, 1761

Herdendorf, Charles E., 'Captain James Cook and the Transits of Mercury and Venus', *The Journal of Pacific History*, 1986, vol. 21

Heyman, Harald J., *Fredrik Mallet och Johan Henrik Lidén. En brevväxling från åren 1769–70*, Uppsala, Lychnos, 1938

Hindle, Brooke, *David Rittenhouse*, Princeton, New Jersey, Princeton University Press, 1964

——*The Pursuit of Science in Revolutionary America 1735–1789*, Chapel Hill, University of North Carolina Press, 1956

Hirst, William, 'Account of Several Phaenomena Observed during the Ingress of Venus into the Solar Disc', *Phil Trans*, 1769, vol. 59

——'An Account of an Observation of the Transit of Venus over the Sun, on the 6th of June 1761, at Madrass', *Phil Trans*, 1761–62, vol. 52, p.397

Holland, Samuel, 'Astronomical Observations Made by Samuel Holland, Esquire, Surveyor-General of Lands for the Northern District of North-America; and Others of His Party', *Phil Trans*, 1769, vol. 59

Home, R. W., 'Science as a Career in Eighteenth-Century Russia: The Case of F. U. T. Aepinus', *The Slavonic and East European Review*, 1973, vol. 51

Hornsby, Thomas, 'A Discourse on the Parallax of the Sun', *Phil Trans*, 1763, vol. 53

——'An Account of the Observations of the Transit of Venus and of the Eclipse of the Sun, Made at Shirburn Castle and at Oxford', *Phil Trans*, 1769, vol. 59

——'On the Transit of Venus in 1769', *Phil Trans*, 1765, vol. 55

——'The Quantity of the Sun's Parallax, as Deduced from the Observations of the Transit of Venus, on June 3, 1769', *Phil Trans*, 1771, vol. 61

Horsfall, James, 'Observation of the Late Transit of Venus', *Phil Trans*, 1769, vol. 59

Horsley, Samuel, 'Venus Observed upon the Sun at Oxford, June 3, 1769', *Phil Trans*, 1769, vol. 59

Howse, Derek, *Francis Place and the Early History of the Greenwich Observatory*, New York, Science History Publications, 1975

——*Nevil Maskelyne: The Seaman's Astronomer*, Cambridge and New York, Cambridge University Press, 1989

Hulshoff Pol, E., 'Een Zweed te Leiden in 1769. Uit het reisdagboek van J.H. Lidén', *Jaarboekje voor Geschiedenis en Oudheidkunde van Leiden en Omstreken*, 1958

Huntington, W. Chapin, 'Michael Lomonosov and Benjamin Franklin: Two Self-Made Men of the Eighteenth Century', *Russian Review*, 1959, vol. 18

James, Lawrence, *The Rise and Fall of the British Empire*, London, Abacus, 2001

Jardine, Alexander, 'Observation of the Transit of Venus, and Other Astronomical Observations, Made at Gibraltar', *Phil Trans*, 1769, vol. 59

Jardine, Lisa, *Ingenious Pursuits: Building the Scientific Revolution*, London, Little, Brown, 1999

Jefferson, Thomas, *Notes on the State of Virginia* (edited by William Peden), New York and London, W. W. Norton, 1982

Jones, Colin, *Paris: Biography of a City*, London, Penguin, 2006

Juskevic, A. P. and E. Winter (ed.), *Die Berliner und die Petersburger Akademie der Wissenschaften im Briefwechsel Leonhard Eulers*, Berlin, Akademie-Verlag, 1959–1976

Kant, Immanuel, *Allgemeine Naturgeschichte und Theorie des Himmels*, Königsberg und Leipzig, Johann Friederich Petersen, 1755

Kästner, I. (ed.), *Wissenschaftskommunikation in Europa im 18. und 19. Jahrhundert*, Aachen, Shaker Verlag, 2009

Kaye, I., 'James Cook and the Royal Society', *Notes and Records of the Royal Society of London*, 1969, vol. 24

Kindersley, Jemima, *Letters from the Island of Teneriffe, Brazil, the Cape of Good Hope, and the East-Indies*, London, 1777

Kordenbusch, Georg Friedrich, *Die Bestimmung der denkwürdigen Durchgänge der Venus durch die Sonne, der Jahre 1761 den 6 Junii und 1769 den 3 Junii*, Nuremberg, 1769

Krafft, Wolfgang Ludwig, *Auszug aus den Beobachtungen welche zu Orenburg bey Gelegenheit des Durgangs der Venus vorbey der Sonnenscheibe angestellt worden sind*, St Petersburg, Imperial Academy of Sciences, 1769

Lalande Jérôme de, 'Explication d'une carte du passage de Vénus sur le disque du Soleil pour le 3 juin 1769', *Histoire & Mémoires*, 1764

——'Extract of a Letter from M. De la Lande, of the Royal Academy of

Sciences at Paris, to the Rev. Mr Nevil Maskelyne, F.R.S. Dated Paris, Nov. 18, 1762', *Phil Trans*, 1761–62, vol. 52

——*Figure du passage de Venus sur le disque du Soleil qu'on observera le 3 juin 1769*, Paris, Jean Lattré, 1760

——'Mémoire sur le passages de Vénus devant le disque du Soleil, en 1761 et 1769', *Histoire & Mémoires*, 1757

——'Observation du passage de Vénus sur le disque du Soleil, faite à Paris au Palais du Luxembourg le 6 Juin 1761', *Histoire & Mémoires*, 1761

——'Observation du passage de Vénus sur le Soleil, faite à Paris le 3 Juin 1769, dans l'Observatoire du Collège Mazarin', *Histoire & Mémoires*, 1769

——'Observations of the Transit of Venus on June 3, 1769, and the Eclipse of the Sun on the Following Day, Made at Paris, and Other Places', *Phil Trans*, 1769, vol. 59

——'Remarques sur les Observations du passage de Vénus, faites à Copenhague & à Drontheim en Norwège, par ordre du roi de Danemarck', *Histoire & Mémoires*, 1761

——'Remarques sur le passage de Vénus, qui s'observera en 1769', *Histoire & Mémoires*, 1768

Le Gentil, Guillaume, *Le Gentils Reisen in den indischen Meeren in den Jahren 1761 bis 1769 und Chappe d'Auteroche Reise nach Mexico und Californien im Jahre 1769 aus dem Französischen, nebst Karl Millers Nachricht von Sumatra und Franziscus Masons Beschreibung der Insel St Miguel aus dem Englischen*, translated by J. P. Ebeling, Hamburg, Carl Ernst Bohn, 1781

——'Mémoire [. . .] au sujet de l'observation qu'il va faire, par ordre du Roi, dans les Indes Orientales, du prochain passage de Vénus pardevant le Soleil', *Le Journal des Sçavans*, 1760

——*Voyage dans les mers de l'Inde*, Paris, Académie des Sciences, 1779 and 1781

Le Monnier, Pierre Charles, 'Observation du passage de Vénus sur le disque du Soleil, faite au château de Saint-Hubert en présence du Roi', *Histoire & Mémoires*, 1761

Leeds, John, 'Observation of the Transit of Venus, on June 3, 1769', *Phil Trans*, 1769, vol. 59

Lentin, A. (ed.), *Voltaire and Catherine the Great. Selected Correspondence*, Cambridge, Oriental Research Partners, 1974

Levitt, Marcus C., 'An Antidote to Nervous Juice: Catherine the Great's Debate with Chappe d'Auteroche over Russian Culture', *Eighteenth Century Studies*, 1998, vol. 32

Lewis, W. S. (ed.), *Horace Walpole's Correspondence*, New Haven and London, Yale University Press, 1937–61

Lexell, Anders, *Disquisitio de investiganda vera quantitate parallaxeos solis ex transitu veneris ante discum solis*, St Petersburg, Imperial Academy of Sciences, 1772

Lincoln, Margarette, (ed.), *Science and Exploration in the Pacific*, London, National Maritime Museum, 1998

Lind, James, 'An Account of the Late Transit of Venus, Observed at Hawkhill, Near Edinburgh', *Phil Trans*, 1769, vol. 59

Lindroth, Sten (ed.), *Swedish Men of Science, 1650–1950*, Stockholm, Swedish Institute, 1952

——*Kungliga Svenska Vetenskapsakademiens Historia*, Stockholm, Almqvist & Wiksell, 1967

Lipski, Alexander, 'The Foundation of the Russian Academy of Sciences', *Isis*, 1953, vol. 44

Littrow, Carl Ludwig (ed.), P. *Hell's Reise nach Wardoe bei Lappland und seine Beobachtung des Venus-Durchganges im Jahre 1769*, Vienna, Carl Gerold Verlag, 1835

Longford, Paul, *A Polite and Commercial People, England 1727–1783*, Oxford and New York, Oxford University Press, 1992

Lowitz, Georg Moritz, *Auszug aus den Beobachtungen welche zu Gurief bey Gelegenheit des Durchgangs der Venus vorbey der Sonnenscheibe angestellt worden sind*, St Petersburg, Imperial Academy of Sciences, 1770

Ludlam, Mr, 'Observations Made at Leicester on the Transit of Venus Over the Sun, June 3, 1769', *Phil Trans*, 1769, vol. 59

MacLeod, Roy (ed.), *Nature and Empire: Science and the Colonial Enterprise*, Chicago, University of Chicago Press, 2000

Madariaga, Isabel de, *Catherine the Great: A Short History*, New Haven and London, Yale University Press, 1990

Magee, William, 'Minutes of the Observation of the Transit of Venus over the Sun, the 6th of June 1761, Taken at Calcutta in Bengal', *Phil Trans*, 1761–62, vol. 52

Mallet, Fredrik, 'De Veneris Transitu, per discum Solis, A. 1761, d. 6 Junii', *Phil Trans*, 1766, vol. 56

——'Nachricht was man bey der Venus Durchgange durch die Sonne den 3. Und 4. Jun. 1769 zu Pello hat beobachten können', *Der Königl. Schwedischen Akademie der Wissenschaften Abhandlungen aus der Naturlehre, Haushaltskunst und Mechanik*, 1769

——'Extract of a Letter from Mr Mallet, of Geneva, to Dr Bevis, F.R.S.', *Phil Trans*, 1770, vol. 60

Maor, Eli, *Venus in Transit*, Princeton, New Jersey, Princeton University Press, 2004

Maraldi, M., 'Observation de la sortie de Vénus du disque du Soleil, faite à l'Observatoire royal le 6 Juin 1761, au matin', *Histoire & Mémoires*, 1761

Mare, Margaret and W. H. Quarrell, (trans. and eds.), *Lichtenberg's Visits to England as Described in his Letters and Diaries*, Oxford, Clarendon Press, 1938

Martin, Benjamin, *Institutions of Astronomical Calculations; Containing a Survey of the Solar System . . . With a Description of Two New Pieces of Mechanism, etc.*, London, the author, 1773

——*The Young Gentleman and Lady's Philosophy, in a Continued Survey of the Works of Nature and Art*, London, W. Owen & the author, 1759

——*Venus in the Sun*, London, 1761

Maskelyne, Nevil, 'A Letter from Revd. Nevil Maskelyne, B.D.F.R.S. Astronomer Royal, to Rev. William Smith, D.D. Provost of the College of Philadelphia, Giving Some Account of the Hudson's-Bay and Other Northern Observations of the Transit of Venus, June 3d, 1769', *Transactions APS*, 1769–71, vol. 1

——'A Letter from the Rev. Nevil Maskelyne, M.A.F.R.S. to William Watson, M.D.F.R.S.', *Phil Trans*, 1761–62, vol. 52

——'An Account of the Observations Made on the Transit of Venus, June 6, 1761, in the Island of St Helena', *Phil Trans*, 1761–62, vol. 52

——*Instructions Relative to the Observation of the Ensuing Transit of the Planet Venus over the Sun's Disk on the 3d of June 1769*, London, Richardson, 1768

——'Observations of the Transit of Venus Over the Sun, and the Eclipse of the Sun, on June 3, 1769; Made at the Royal Observatory', *Phil Trans*, 1768, vol. 58

Mason, Charles and Jeremiah Dixon, 'Observations Made at the Cape of Good Hope', *Phil Trans*, 1761–62, vol. 52

Mason, Charles, 'Astronomical Observations Made at Cavan, Near Strabane, in the County of Donegal, Ireland, by Appointment of the Royal Society', *Phil Trans*, 1770, vol. 60

Mason, Hughlett, 'The Journal of Charles Mason and Jeremiah Dixon, 1763–1768', *Memoirs of the APS*, 1969, vol. 76

Masserano, Prince, 'A Short Account of the Observations of the Late Transit of Venus, Made in California, by Order of His Catholic Majesty; Communicated by His Excellency Prince Masserano, Ambassador from the Spanish Court', *Phil Trans*, 1770, vol. 60

Mayer, Andreas, 'Observatio Ingressus Veneris in Solem 3 die Junii, 1769, habita Gryphiswaldiae', *Phil Trans*, 1769, vol. 59

Mayer, Christian, *Ad augustissimam Russiarum omnium imperatricem Catharinam II. Alexiewnam expositio de transitu Veneris ante discum Solis d. 23 Maii, 1769*, St Petersburg, Imperial Academy of Sciences, 1769

——'An Account of the Transit of Venus', *Phil Trans*, 1764, vol. 54

——'Expositio utriusque observationes et Veneris et eclipsis Solaris', *Novi Commentarii Academi Scientiarum Imperialis Petropolitan, 1768*, vol. 13 and 1769, vol. 14

McClellan, James E., *Science Reorganized. Scientific Societies in the Eighteenth Century*, New York, Columbia University Press, 1985

McClellan, James E. and François Regourd, 'The Colonial Machine: French Science and Colonisation in the Ancien Regime', in *Nature and Empire: Science and the Colonial Enterprise* by Roy MacLeod (ed.), Chicago, University of Chicago Press, 2000

McLynn, Frank, *Captain Cook*, New Haven and London, Yale University Press, 2011

Meadows, A. J., 'The Discovery of an Atmosphere on Venus', *Annals of Science*, 1966, vol. 22

Melander, Daniel, 'Erklärung der Erscheinungen die sich bey der Venus Durchgange durch die Sonne zeigen', *Der Königl. Schwedischen Akademie der Wissenschaften Abhandlungen aus der Naturlehre, Haushaltskunst und Mechanik*, 1769

Menshutkin, Boris Nikolaevich, *Russia's Lomonosov: Chemist, Courtier, Physicist, Poet*, Oxford, Oxford University Press, 1952

Millburn, John R., Benjamin Martin: *Author, Instrument-maker, and Country Showman*, Leyden, Noordhoff International Publications, 1976

Miller, David Philip and Peter Hanns Reill (eds.), *Visions of Empire: Voyages, Botany, and Representations of Nature*, Cambridge, Cambridge University Press, 1991

Mohr, Johan Maurits, 'Transitus Veneris & Mercurii in Eorum Exitu è Disco Solis, 4to Mensis Junii & 10mo Novembris, 1769, Observatus. Communicated by Capt. James Cook', *Phil Trans*, 1771, vol. 61

Morosow, A. A., *Michail Wassiljewitsch Lomonosow*, Berlin, Rütten & Loening, 1954

Morris, Margaret, 'Man Without A Face – Charles Green', *Cook's Log*, 1980, vol. 3 and 1981, vol. 4

Moutchnik, Alexander, *Forschung und Lehre in der zweiten Hälfte des 18. Jahrhunderts: der Naturwissenschaftler und Universitätsprofessor Christian Mayer SJ (1719–1783)*, Augsburg, E. Rauner, 2006

Munck, Thomas, *The Enlightenment. A Comparative Social History 1721–1794*, London, Hodder, 2000

Muyden, Madame van (trans. and ed.), *A Foreign View of England in the Reigns of George I & George II. The Letters of Monsieur César de Saussure to his Family*, London, John Murray, 1902

Mylius, Christlob, 'Christlob's Mylius Tagebuch seiner Reise von Berlin nach England, in Bernoulli, Johann, *Archiv zur neuern Geschichte, Geographie, Natur – und Menschenkenntnis*, vol. 7, Leipzig, Georg Emanuel Beer, 1787

Nettel, Reginald (trans. and ed.), *Carl Philip Moritz. Journeys of a German in England in 1782*, London, Jonathan Cape, 1965

Nevskaia, Nina Ivanovna, *Joseph-Nicolas Delisle (1688–1768)*, Paris, 1973

Newcomb, Simon, 'Discussion of Observations of the Transits of Venus in 1761 and 1769', *Astronomical Papers prepared for the use of the American Ephemeris and Nautical Almanac*, 1890, vol. 2

Nordenmark, N. V. E., *Fredrik Mallet och Daniel Melanderhjelm två Uppsala-Astronomer*, Uppsala, Almqvist & Wiksells, 1946

——*Pehr Wilhelm Wargentin: Kungl. Vetenskapsakademiens Sekreterare och Astronom, 1749–1783*, Uppsala, Almqvist & Wirsells, 1939

Nunis, Doyce B., *The 1769 Transit of Venus: The Baja California Observations of Jean-Baptiste Chappe d'Auteroche, Vicente de Doz, and Joaquín Velázquez Cárdenas de León*, Los Angeles, Natural History Museum of Los Angeles County, 1982

Parkinson, Sidney, *A Journal of a Voyage to the South Seas, in His Majesty's ship, the* Endeavour, London, 1784

Pauly, Philip P., Fruits and Plains: *The Horticultural Transformation of America*, Cambridge, Hassachusetts, Harvard University Press, 2007

Pekarsky, P., *Istoriya Imperartorskoy Akademii nuak v Peterburge*, St Petersburg, 1870–73

Peynson, Lewis and Susan Sheets-Peynson, *Servants of Nature. The*

History of Scientific Institutions, Enterprises and Sensibilities, London, HarperCollins, 1999

Pfrepper, Regine and Gerd Pfrepper, 'Georg Moritz Lowitz (1722–1774) und Johann Tobias Lowitz (1757–1804) – zwei Wissenschaftler zwischen Göttingen und St Petersburg', in Mittler, Elmar and Silke Glitsch (ed.), *300 Jahre St Petersburg. Russland und die 'Göttingische Seele'*, Göttingen, Niedersächsische Staats-und Universitätsbibliothek, 2004

——'Georg Moritz Lowitz (1722–1774). Astronom und Geograph im Auftrag der St Petersburger Akademie der Wissenschaften', in Kästner, I. (ed.), *Wissenschaftskommunikation in Europa im 18. und 19. Jahrhundert*, Aachen, Shaker Verlag, 2009

Pigatto, Louisa, 'The 1761 Transit of Venus Dispute between Audiffredi and Pingré', in Kurtz, D. W. (ed.), *Transits of Venus: New Views of the Solar System and Galaxy*, Proceedings IAU Colloquium, 2004, no. 196

Pigott, Nathan, 'On the Late Transit of Venus', *Phil Trans*, 1770, vol. 60

Pingré, Alexandre-Gui, 'A Letter from M. Pingré, of the Royal Academy of Sciences at Paris, to the Rev. Mr Maskelyne, Astronomer Royal, F.R.S', *Phil Trans*, 1770, vol. 60

——'A Supplement to Mons. Pingré's Memoir on the Parallax of the Sun', *Phil Trans*, 1764a, vol. 54

——'Mémoire sur le choix et l'état des lieux où le passage de Vénus, du 3 Juin 1769, pourra être observé avec le plus d'avantage et principalement sur la position géographique des isles de la mer du Sud', *Histoire & Mémoires*, 1767

——'Mémoire sur l'observation du passage de Vénus sur le disque du soleil, faite à Séleninsk en Sibérie', *Histoire & Mémoires*, 1764b

——'Nouvelle recherche sur la détermination de la parallaxe du Soleil par le passage de Vénus du 6 Juin 1761', *Histoire & Mémoires*, 1765

——'Observation du passage de Vénus, sur le disque du Soleil, faite au Cap François, isle de Saint-Domigue, le 3 Juin 1769', *Histoire & Mémoires*, 1769

——'Observation du passage de Vénus sur le disque du Soleil, le 6 Juin 1761, faite à Rodrigue dans la Mer des Indes', *Histoire & Mémoires*, 1761a

——'Observation of the Transit of Venus over the Sun, June 6, 1761, at the Island of Rodrigues', *Phil Trans*, 1761–62, vol. 52

——'Observations astronomiques pour la determination de la parallaxe du Soleil, faites en l'isle Rodrigue', *Histoire & Memoires*, 1761b

——'Précis d'un voyage en Amérique', *Histoire & Mémoires*, 1770

——*Voyage à Rodrigue. Le Transit de Vénus de 1761, la Mission Astronomique de L'Abbé Pingré dans l'Océan Indien* (Sophie Hoarau, Marie-Paul Janiç and Jean-Michel Racault, eds.), Saint-Denis, Université de la Réunion, 2004

Planman, Anders, 'A Determination of the Solar Parallax Attempted, by a Peculiar Method, from the Observations of the Last Transit of Venus', *Phil Trans*, 1768, vol. 58

——'An Account of the Observations Made upon the Transit of Venus over the Sun, 6th June 1761, at Cajaneburg in Sweden, by Mons. Planman', *Phil Trans*, 1761–62, vol. 52

——'Die Parallaxe der Sonne', *Der Königl. Schwedischen Akademie der Wissenschaften Abhandlungen aus der Naturlehre, Haushaltskunst und Mechanik*, 1763

——'Geographische Lage von Cajaneborg', *Der Königl. Schwedischen Akademie der Wissenschaften Abhandlungen aus der Naturlehre, Haushaltskunst und Mechanik*, 1762

——'Geographische Lage von Cajaneborg', *Der Königl. Schwedischen Akademie der Wissenschaften Abhandlungen aus der Naturlehre, Haushaltskunst und Mechanik*, 1768

——'Venus in der Sonne, den 3. Jun. 1769 beobachted zu Cajaneborg', *Der Königl. Schwedischen Akademie der Wissenschaften Abhandlungen aus der Naturlehre, Haushaltskunst und Mechanik*, 1769

Porter, James, 'Observations on the Same Transit of Venus Made at Constantinople', *Phil Trans*, 1761–62, vol. 52

Porter, Roy, *English Society in the Eighteenth Century*, London, Penguin, 1990

——*Enlightenment. Britain and the Creation of the Modern World*, London, Penguin, 2001

——*The Cambridge History of Science*, Cambridge, Cambridge University Press, 2003, vol. 4

Proctor, Richard A., *Transits of Venus. A Popular Account of Past and Coming Transits*, London, Longmans, Green and Co., 1882

Prosperin, Erik, 'Auszug aus den Beobachtungen des Eintritts der Venus in die Sonne den 3ten Jun. 1769, welche auf der Sternwarte zu Upsala gehalten worden', *Der Königl. Schwedischen Akademie der Wissenschaften Abhandlungen aus der Naturlehre, Haushaltskunst und Mechanik*, 1769

Putnam, Peter (ed.), *Seven Britons in Imperial Russia 1698–1812*, Princeton, New Jersey, Princeton University Press, 1952

Ratcliff, Jessica, *The Transit of Venus Enterprise in Victorian Britain*, London, Pickering & Chatto, 2008

Richardson, *Longitude and Empire: How Captain Cook's Voyages Changed the World*, Vancouver, University of British Columbia Press, 2005

Rittenhouse, David, 'Calculations of the Transit of Venus over the Sun as It Is to Happen June 3d 1769, in lat 40°. N. Long. 5h West from Greenwich', *Transactions APS*, 1769–71, vol. 1

Robinson, H. W., 'A Note on Charles Mason's Ancestry and His Family', *Proceedings APS*, 1949, vol. 93

——'Jeremiah Dixon (1733–1779): A Biographical Note', *Proceedings APS*, 1950, vol. 94

Röhl, Lampert Hinrich, *Merkwürdigkeiten von den Durchgängen der Venus durch die Sonne*, Greifswald, A. F. Röse, 1768

Ronan, Colin A., *Edmond Halley: Genius in Eclipse*, London, Macdonald & Co., 1970

Rose, Alexander, 'Extract of Two Letters from the Late Capt. Alexander Rose, of the 52d Regiment, to Dr Murdoch, F.R.S.', *Phil Trans*, 1770, vol. 60

Rounding, Virginia, *Catherine the Great. Love, Sex and Power*, London, Arrow Books, 2007

Rufus, Carl W., 'David Rittenhouse, Pioneer American Astronomer', *The Scientific Monthly*, 1928, vol. 26

Rush, Benjamin, *Eulogium Intended to Perpetuate the Memory of David Rittenhouse*, Philadelphia, J. Omrod, 1796

Sarton, George, 'Vindication of Father Hell', *Isis*, 1944, vol. 35, no. 2

Sawyer Hogg, Helen, 'Out of Old Books', *Journal of the Royal Astronomical Society of Canada*, 1951, vol. 45

Schenmark, Nils, 'Beobachtungen des Eintritts der Venus in die Sonne, den 3.Jun. und der Sonnenfinsterniß den 4. Jun. dieses Jahrs, angestellt zu Lund', *Der Königl. Schwedischen Akademie der Wissenschaften Abhandlungen aus der Naturlehre, Haushaltskunst und Mechanik*, 1769

Schulze, Ludmilla, 'The Russification of the St Petersburg Academy of Sciences and Arts in the Eighteenth Century', *The British Journal for the History of Science*, 1985, vol. 18

Scott, Robert Henry, 'The History of the Kew Observatory', *Proceedings of the Royal Society of London*, 1885, vol. 39

Sheehan, William and John Westfall, *The Transits of Venus*, Amherst, Prometheus Books, 2004

Short, James, 'An Account of the Transit of Venus over the Sun, on Saturday Morning, 6th June 1761, at Savile-House', *Phil Trans*, 1761–62, vol. 52

——'Second Paper concerning the Parallax of the Sun Determined from the Observations of the Late Transit of Venus', *Phil Trans*, 1763, vol. 53

——'The Observations of the Internal Contact of Venus with the Sun's Limb, in the Late Transit, Made in Different Places of Europe, Compared with the Time of the Same Contact Observed at the Cape of Good Hope, and the Parallax of the Sun from Thence Determined', *Phil Trans*, 1761–62, vol. 52

Smith, Edwin Burrows, 'Jean-Sylvain Bailly: Astronomer, Mystic, Revolutionary 1736–1793', *Transactions APS*, New Series, 1954, vol. 44

Smith, James Edward (ed.), *A Selection of the Correspondence of Linnaeus and other Naturalists*, London, Longman, 1821

Smith, William, John Ewing, Owen Biddle, Hugh Williamson, Thomas Combe, and David Rittenhouse, 'Apparent Time of the Contacts of the Limbs of the Sun and Venus; With Other Circumstances of Most Note, in the Different European Observations of the Transit, June 3d, 1769', *Transactions APS*, 1769–71, vol. 1

Smith, William, John Lukens, David Rittenhouse and John Sellers, 'Account of the Transit of Venus Over the Sun's Disk, as Observed at Norriton, in the County of Philadelphia, and Province of Pennsylvania, June 3, 1769', *Phil Trans*, 1769, vol. 59

——'An Account of the Transit of Venus over the Sun, June 3d, 1769, as Observed at Norriton, in Pennsylvania', *Transactions APS*, 1769–71, vol. 1

Sobolevskii, S. A., *Kamer-fur'erskie zhurnaly 1695–1774*, Moscow, T. Ris, 1853–67 (Catherine's Court Journals)

Sörlin, Sverker, 'Ordering the World for Europe: Science as Intelligence and Information as Seen from the Northern Periphery', in *Nature and Empire: Science and the Colonial Enterprise* by Roy MacLeod (ed.), Chicago, University of Chicago Press, 2000

Spindler, Max (ed.), *Electoralis Academiae Scientiarum Boicae primordia. Briefe aus der Gründungszeit der Bayerischen Akademie der Wissenschaften*, München, C. H. Bck'sche Verlagsbuchhandlung, 1959

Swift, Jonathan, *Gulliver's Travels*, London, Jones & Co., 1826

Teets, Donald A., 'Transits of Venus and the Astronomical Unit', *Mathematics Magazine*, 2003, vol. 76

Turner, G. L. E., 'James Short, F.R.S., and His Contribution to the Construction of Reflecting Telescopes', *Notes and Records of the Royal Society of London*, 1969, vol. 24

Uglow, Jenny, *The Lunar Men: The Friends Who Made the Future*, London, Faber and Faber, 2002

Veselovsky, Konstantin Stepanovich (ed.), *Protokoly zasedaniy konferentsii Imperatorskoy Akademii nauk s 1725 po 1803 goda*, St Petersburg, Imperial Academy of Sciences, 1897–1911

Vucinich, Alexander, *Empire of Knowledge: The Academy of Sciences of the USSR (1917–1970)*, Berkeley, University of California Press, 1984

Wales, Wendy, 'William Wales' First Voyage', *Cook's Log*, 2004, vol. 27

Wales, William and Joseph Dymond, 'Astronomical Observations Made by Order of the Royal Society, at Prince of Wales's Fort, on the North-West Coast of Hudson's Bay', *Phil Trans*, 1769, vol. 59

——'Observations on the State of the Air, Winds, Weather, &c. Made at Prince of Wales's Fort, on the North-West Coast of Hudson's Bay, in the Years 1768 and 1769', *Phil Trans*, 1770, vol. 60

Wales, William, 'Journal of a Voyage, Made by Order of the Royal Society, to Churchill River, on the North-West Coast of Hudson's Bay, Of Thirteen Months Residence in That Country; and of the Voyage Back to England, In the Years 1768 and 1769', *Phil Trans*, 1770, vol. 60

Wargentin, Pehr Wilhelm, 'A Letter from Monsieur Wargentin, Secretary to the Royal Academy of Sciences in Sweden, to Mr John Ellicott, F.R.S. Relating to the Late Transit of Venus', *Phil Trans*, 1763, vol. 53

——'An Account of the Observations Made on the Same Transit in Sweden', *Phil Trans*, 1761–62, vol. 52

——'Anmerkungen über den Durchgang der Venus durch die Sonnenscheibe', *Der Königl. Schwedischen Akademie der Wissenschaften Abhandlungen aus der Naturlehre, Haushaltskunst und Mechanik*, 1761

——'Beobachtungen der Venus durch die Sonne, den 6 Jun.1761', *Der Königl. Schwedischen Akademie der Wissenschaften Abhandlungen aus der Naturlehre, Haushaltskunst und Mechanik*, 1761

——'Bericht von den Anstalten, die in Schweden sind gemacht worden, den 3 Jun. 1769, zu beobachten, und wie solche gelungen sind; nebst den Stockholmischen Beobachtungen', *Der Königl. Schwedischen Akademie der Wissenschaften Abhandlungen aus der Naturlehre, Haushaltskunst und Mechanik*, 1769

——'Von dem Unterschiede der Mittagstreife der Oerter da Venus den 6 Jun. 1761 in der Sonne beobachted worden ist', *Der Königl. Schwedischen Akademie der Wissenschaften Abhandlungen aus der Naturlehre, Haushaltskunst und Mechanik*, 1763

Watlington, Hereward T., *Family Narrative*, Devonshire, Bermuda, privately printed, 1980

Weigley, Russell F. (ed.), *Philadelphia. A 300-Year History*, New York, W. W. Norton, 1982

Wendland, Folkwart, *Peter Simon Pallas (1741–1811): Materialien einer Biographie*, Berlin and New York, Walter de Gruyter, 1992

West, Benjamin, 'An Account of the Transit of Venus over the Sun, June 3rd, 1769, as Observed at Providence, New England', *Transactions APS*, 1769–71, vol. 1

Westenrieder, Ludwig, *Geschichte der Baierischen Akademie der Wissenschaften*, Munich, Akademischer Bücherverlage, 1784

Widmalm, Sven, 'A Commerce of Letters: Astronomical Communication in the 18th Century', *Science Studies*, 1992, vol. 5

Williams, Samuel, 'An Account of the Transit of Venus over the Sun, June 3rd, 1769, as Observed at Newbury in Massachusetts', *Transactions APS*, 1786, vol. 2

Wilson, Alexander, 'Observations of the Transit of Venus Over the Sun', *Phil Trans*, 1769, vol. 59

Winthrop, John, 'Observation of the Transit of Venus, June 6, 1761, at St John's Newfound-Land', *Phil Trans*, 1764, vol. 54

——'Observations of the Transit of Venus Over the Sun, June 3, 1769', *Phil Trans*, 1769, vol. 59

——*Relation of a Voyage from Boston to Newfoundland, for the Observation of the Transit of Venus, June 6, 1761*, Boston, Massachusetts, Edes & Gill, 1761

——*Two Lectures on the Parallax and Distance of the Sun, as Deducible from the Transit of Venus*, Boston, Massachusetts, Edes & Gill, 1769b

Wolff, Larry, *Inventing Eastern Europe: The Map of Civilization on the Mind of the Enlightenment*, Stanford, California, Stanford University Press, 1994

Wollaston, Francis, 'Observations of the Transit of Venus Over the Sun, on June 3, 1769; and the Eclipse of the Sun the Next Morning; Made at East Dereham, in Norfolk', *Phil Trans*, 1769, vol. 59

Woolf, Harry, *The Transits of Venus. A Study of Eighteenth Century Science*, Princeton, New Jersey, Princeton University Press, 1959

Woolley, Richard, 'Captain Cook and the Transit of Venus of 1769', *Notes and Records of the Royal Society of London*, 1969, vol. 24

Wright, Thomas, 'An Account of an Observation of the Transit of Venus, Made at Isle Coudre Near Quebec', *Phil Trans*, 1769, vol. 59

Wulf, Andrea, *The Brother Gardeners. Botany, Empire and the Birth of an Obsession*, London, William Heinemann, 2008

Zanotti, Eustachio, 'De Veneris ac Solis Congressu Observatio, Habita in Astronomica Specula Bononiensis Scientiarum Instituti, Die 5 Junii 1761', *Phil Trans*, 1761–62, vol. 52

Zuidervaart, Huib J. and Rob H. Van Gent, '"A Bare Outpost of Learned European Culture on the Edge of the Jungles of Java": Johan Maurits Mohr (1716–1775) and the Emergence of Instrumental and Institutional Science in Dutch Colonial Indonesia', *Isis*, 2004, vol. 95

Suggested Further Reading

Transit 2012

The next and last transit in our lifetime will be on 5 and 6 June 2012 (depending where you are).

If you want to watch the transit and you're trying to work out where best to go, or if you need a transit calculator, a parallax calculator, information on how to protect your eyes, the science behind the transit and so on, there are some excellent websites which also include modern mappemondes.

For example:

www.transitofvenus.nl
www.transitofvenus.org

William Sheehan's and John Westfall's book *The Transits of Venus* (2004) also includes useful tables of 'Local Circumstances' and 'Sunshine Probability Maps'.

Scientific Information:

For more detailed information about the solar parallax and the calculations of the transits of Venus:

Teets, Donald A., 'Transits of Venus and the Astronomical Unit', *Mathematics Magazine*, 2003, vol. 76,

Woolf, Harry, *The Transits of Venus. A Study of Eighteenth Century Science*, Princeton, New Jersey, Princeton University Press, 1959

Venustransit 2004 (by Heinz Blatter)
http://eclipse.astroinfo.org/transit/venus/project2004/pub/Blatter.etal.eng.200306.pdf

The Transit of Venus, Workbook (by Steven van Roode)
http://www.transitofvenus.nl/files/TransitOfVenus.pdf

The Transit of Venus & The Quest for the Solar Parallax (by David Sellers)
http://homepage.ntlworld.com/magavelda/ds/venus/ven_ch1A.htm

From Stargazers to Starships (by David P. Stearn)
http://www.-istp.gsfc.nasa.gov/stargaze/Svenus1.htm

Picture Credits

Plate Section
Edmond Halley, line engraving after R. Phillips, n.d. Reproduced with the permission of the Wellcome Library, London.

Joseph-Nicolas Delisle. Reproduced with the permission of Bibliothèque de l'Observatoire de Paris.

Alexandre-Gui Pingré, bust by Jean-Jacques Caffieri. Courtesy of the Bibliothèque Sainte-Geneviève.

Nevil Maskelyne, stipple engraving by Edward Scriven after van der Burgh, n.d. Reproduced with the permission of the Wellcome Library, London.

Jean-Baptiste Chappe d'Auteroche, line engraving by J. B. Tilliard, 1772, after J. M. Frédou. Reproduced with the permission of the Wellcome Library, London.

Mikhail Lomonosov, colour reproduction of painting, 1953, after Steiner. Reproduced with the permission of the Wellcome Library, London.

Pehr Wilhelm Wargentin, in *Svenska Familj-Journalen*, 1879, vol. 18.

Catherine the Great, line engraving by J. Miller, n.d. Reproduced with the permission of the Wellcome Library, London.

James Cook, line engraving by J. K. Sherwin, 1779, after Sir N. Dance-Holland, 1776. Reproduced with the permission of the Wellcome Library, London.

Maximilian Hell, portrait by W. Pohl and engraved in Vienna in 1771. Reproduced with the permission of the Universitätssternwarte Wien.

Benjamin Franklin, engraving by E. Savage after a painting by David Martin, 1767. Reproduced with the permission of the Wellcome Library, London.

David Rittenhouse, after Charles Willson Peale, c.1791–96. Reproduced with the permission of the Independence National Historical Park.

The Royal Observatory, Paris: the terrace on the garden side, with men experimenting with astronomical and other scientific instruments, etching, early eighteenth century, after Claude Perrault. Reproduced with the permission of the Wellcome Library, London.

Regulator clock, 1768–69, made by John Shelton. Reproduced with the permission of the Science Museum/SSPL, London.

Gregorian reflecting telescope, c.1760, made by James Short. Reproduced with the permission of the Science Museum/SSPL, London.

Twelve-inch portable astronomical quadrant, 1760–69, made by John Bird. Reproduced with the permission of the Science Museum/SSPL, London.

Images in Text

All maps and scientific illustrations by John Gilkes

Prologue: Edmond Halley, 'Methodus singularis quâ Solis Parallaxis sive distantia à Terra, ope Veneris intra Solem conspiciendæ, tuto determinari poterit', *Phil Trans*, 1714–16. Reproduced with the permission of the Wellcome Library, London.

Prologue: Benjamin Martin, *The Young Gentleman and Lady's Philosophy, in a Continued Survey of the Works of Nature and Art*, London, W. Owen & the author, 1759. Reproduced with the permission of the Wellcome Library, London.

Chapter 1: Mappemonde, in James Ferguson, *Astronomy Explained upon Sir Isaac Newton's Principles*, London, W. Strahan, 1770. Reproduced with the permission of the Wellcome Library, London.

Chapter 1: Royal Society, in Walter Thornbury, *Old and New London*, 1878, vol. 1. Reproduced with the permission of the Wellcome Library, London.

Chapter 2: Detail from Maximilian Hell's map of Vardø, 'Insula Wardoehus', in Maximilian Hell, *Ephemerides Astronomicae ad Meridianum Vindobonensem Anni 1791*. Reproduced with the permission of the Universitätssternwarte Wien.

Chapter 3: The Royal Observatory, Greenwich Hill. Engraving, n.d. Reproduced with the permission of the Wellcome Library, London.

Chapter 3: James Town, St Helena, engraving by William Daniell from the drawings of Samuel Davis, in Alexander Beatson, *Tracts Relative to the Island of St Helena, Written During a Residence of Five Years*, London, printed by W. Bulmer and Co., for G. & W. Nicol & J. Booth, 1816.

Chapter 4: Imperial Academy of Sciences, in A. B., Granville, *St Petersburgh: A Journal of Travels to and from that Capital. Through Flanders, the Rhenich provinces, Prussia, Russia, Poland, Silesia, Saxony, the Federated States of Germany, and France*, London, H. Colburn, 1829.

Chapter 4: Enclosed sledges, in Jean-Baptiste Chappe d'Auteroche, *Voyage en Sibérie, fait par ordre du roi en 1761*, Paris, 1768. Reproduced with the permission of the Wellcome Library, London.

Chapter 4: Russian cottage, in Jean-Baptiste Chappe d'Auteroche, *Voyage en Sibérie, fait par ordre du roi en 1761*, Paris, 1768. Reproduced with the permission of the Wellcome Library, London.

Chapter 5: A large observatory telescope, with an astronomer recording the transit of Venus, engraving by James Basire, eighteenth century. Reproduced with the permission of the Wellcome Library, London.

Chapter 5: Benjamin Martin, *The Young Gentleman and Lady's Philosophy, in a Continued Survey of the Works of Nature and Art*, London, W. Owen & the author, 1759. Reproduced with the permission of the Wellcome Library, London.

Chapter 5: Map of St Helena, engraving by William Daniell from the drawings of Samuel Davis, in Alexander Beatson, *Tracts Relative to the Island of St Helena, Written During a Residence of Five Years*, London, printed by W. Bulmer and Co., for G. & W. Nicol & J. Booth, 1816.

Chapter 5: Cape Town, Cape of Good Hope, in the Diary of Thomas Graham, *c.*1849–50. Reproduced with the permission of the Wellcome Library, London.

Chapter 5: Tobolsk, in Jean-Baptiste Chappe d'Auteroche, *Voyage en Sibérie, fait par ordre du roi en 1761*, Paris, 1768. Reproduced with the permission of the Wellcome Library, London.

Chapter 6: Transit of Venus, in James Ferguson, *Astronomy Explained upon Sir Isaac Newton's Principles*, London, W. Strahan, 1770. Reproduced with the permission of the Wellcome Library, London.

Chapter 6: A large reflecting telescope, and projection of the transit of Venus, engraving, n.d., probably after an engraving in Benjamin Martin, *The General Magazine of Arts and Sciences*, 1755–64. Reproduced with the permission of the Wellcome Library, London.

Chapter 7: Luminous ring and black drop effect, in Torbern Bergman, 'An Account of the Observations Made on the Same Transit at Upsal in Sweden', *Phil Trans*, 1761–62, vol. 52. Reproduced with the permission of the Wellcome Library, London.

Chapter 8: Map of Madagascar, in Guillaume Le Gentil, *Voyage dans les mers de l'Inde*, Paris, Académie des Sciences, Paris, 1779 and 1781, vol. 2. Reproduced with the permission of the Universitätssternwarte Wien.

Chapter 9: Winter Palace in St Petersburg, in A. B. Granville, *St Petersburgh: A Journal of Travels to and from that Capital. Through Flanders, the Rhenich provinces, Prussia, Russia, Poland, Silesia, Saxony, the Federated States of Germany, and France*, London, H. Colburn, 1829.

Chapter 9: Russian gentleman dressed in furs, in William Coxe, *Travels in Poland, Russia, Sweden, and Denmark. Illustrated with charts and engravings*, London, printed for T. Cadell Jr. and W. Davies, 1802.

Chapter 10: A portable observatory tent, engraving by Robert Bénard, after Louis-Jacques Goussier, in D. Diderot and J. le

R. D'Alembert, *Encyclopédie, ou dictionnaire raisonné des sciences, des arts, et des métiers*, Paris, 1762–73. Reproduced with the permission of the Wellcome Library, London.

Chapter 12: A gigantic iceberg at sea, dwarfing a sailing ship, coloured wood engraving by Charles Whymper, S. Bentley and Co., London, n.d. Reproduced with the permission of the Wellcome Library, London.

Chapter 12: 'A North West View of Prince of Wales's Fort in Hudson's Bay, North America', engraving by Samuel Hearne, 1777, Reproduced with the permission of the Hudson's Bay Company Archives, Archives of Manitoba.

Chapter 12: David Rittenhouse's observatory in Norriton, in Theodore W. Bean (ed.), *History of Montgomery County, Pennsylvania*, Philadelphia, Pennsylvania, Everts & Peck, 1884.

Chapter 13: Map of Northern Norway, in William Bayley, 'Astronomical Observations Made at the North Cape, for the Royal Society', *Phil Trans*, 1769, vol. 59. Reproduced with the permission of the Wellcome Library, London.

Chapter 13: Map of Manila to Pondicherry, in Guillaume Le Gentil, *Voyage dans les mers de l'Inde,* Paris, Académie des Sciences, Paris, 1779 and 1781, vol. 1. Reproduced with the permission of the Universitätssternwarte Wien.

Chapter 13: Pondicherry, in Guillaume Le Gentil, *Voyage dans les mers de l'Inde,* Paris, Académie des Sciences, Paris, 1779 and 1781, vol. 1. Reproduced with the permission of the Universitätssternwarte Wien.

Chapter 13: Map from Maximilian Hell's diary of his voyage to Vardø. Reproduced with the permission of the Universitätssternwarte Wien.

Chapter 13 Hell's ship, in Maximilian Hell, *Ephemerides Astronomicae ad Meridianum Vindobonensem Anni 1791*. Reproduced with the permission of the Universitätssternwarte Wien.

Chapter 13: Observatory in Vardø, in Maximilian Hell, Ephemerides Astronomicae ad Meridianum Vindobonensem Anni 1791. Reproduced with the permission of the Universitäts-sternwarte Wien.

Chapter 13: Matavai Bay, in John Hawkesworth, *An Account of the Voyages Undertaken by the Order of His Present Majesty for Making Discoveries in the Southern Hemisphere*, London, W. Strahan and T. Cadell, 1773, vol. 2. Reproduced with the permission of the Wellcome Library, London.

Chapter 13: Fort Venus, in Sydney Parkinson, *A Journal of a Voyage to the South Seas, in His Majesty's ship, the* Endeavour, London, 1784. Reproduced with the permission of the Wellcome Library, London.

Chapter 13: Dancing scene in Tahiti, in John Hawkesworth, *An Account of the Voyages Undertaken by the Order of His Present Majesty for Making Discoveries in the Southern Hemisphere*, London, W. Strahan and T. Cadell, 1773, vol. 2. Reproduced with the permission of the Wellcome Library, London.

Chapter 14: Map of Tahiti, in George William Anderson, *A New, Authentic, and Complete Collection of Voyages Round the World, Undertaken and Performed by Royal Authority*, London, 1800. Reproduced with the permission of the Wellcome Library, London.

Chapter 14: Luminous ring by Cook and Green, in James Cook, 'Observations Made, by Appointment of the Royal Society, at King George's Island in the South Sea; By Mr Charles Green, Formerly Assistant at the Royal Observatory at Greenwich, and Lieut. James Cook, of His Majesty's Ship the *Endeavour*', *Phil Trans*, 1771, vol. 61. Reproduced with the permission of the Wellcome Library, London.

Chapter 14: Map of Vardø, 'Insula Wardoehus', in Maximilian Hell, *Ephemerides Astronomicae ad Meridianum Vindobon-ensem Anni 1791*. Reproduced with the permission of the Universitätssternwarte Wien.

Chapter 14: 'Artificial Transit', Benjamin Martin, *Institutions of Astronomical Calculations Containing a Survey of the Solar System*, London, 1773. Courtesy of the Adler Planetarium & Astronomy Museum, Chicago, Illinois.

Chapter 14: Viewing the Transit of Venus, satirical print, published by Sayer & Co., London, 1793, AN518590001 © Trustees of the British Museum.

Chapter 15: Luminous ring and the black drop effect, in Samuel Dunn, 'A Determination of the Exact Moments of Time When the Planet Venus Was at External and Internal Contact with the Sun's Limb, in the Transits of June 6th, 1761, and June 3d, 1769', *Phil Trans*, 1770, vol. 60. Reproduced with the permission of the Wellcome Library, London.

Chapter 15: Repairs to the *Endeavour*, in George William Anderson, *A New, Authentic, and Complete Collection of Voyages Round the World, Undertaken and Performed by Royal Authority*, London, 1800. Reproduced with the permission of the Wellcome Library, London.

Chapter 15: Tygers Street Canal in Batavia, in *A Collection of Voyages and Travels, Some Now First Printed from Original Manuscripts, Others Now First Published in English*, London, printed by assignment from Messrs. Churchill, 1744–46. Reproduced with the permission of the Wellcome Library, London.

Epilogue: Thomas Hornsby, 'The Quantity of the Sun's Parallax, as Deduced from the Observations of the Transit of Venus, on June 3, 1769', *Phil Trans*, 1771, vol. 61.

Acknowledgements

I have received an incredible amount of support and help from friends, family and strangers alike. Like the preparations for the transits in 1761 and 1769, writing *Chasing Venus* turned out to be an international project as well as a race against a celestial deadline.

Thank you so much to Jo Dunkley for patiently explaining the workings of the transit and the solar parallax to an astronomical novice (all mistakes are entirely my own!). I would like to thank: Alison Boyle and David Rooney at the Science Museum, London; David Butterfield for his Latin translations; Anders Jansson for Swedish research in Stockholm and translations; Oleksandr Karpenko for Russian research and translations; Felix von Reiswitz for French research and translations; Steven van Roode at the Transit of Venus Project for his amazing transit 'calculator' and help, as well as to all the other contributors to the blog; Simon Schaffer; Tofigh Heidarzadeh at the Huntington Library; Regina von Berlepsch at the Astrophysikalisches Institut, Potsdam; Regine Pfrepper; Chris Lintott; Simon Dixon; Connie Wall; Pedro Ferreira; and Dr Jürgen Hamel for his kind help in sending books, articles, pictures and comments on the chapters.

I would also like to thank Elaine Grublin at the Massachusetts Historical Society, Boston; staff at the American Philosophical Society, Philadelphia; Inga Elmqvist at Stockholm Observatory; Anne Miche de Malleray at the Center for History of Science in Stockholm; Keith Moore, Felicity Henderson and the staff at the Royal Society library, London; the staff at the British Library, Wellcome Library and London Library; Gloria Clifton at the Royal Observatory in Greenwich and Rebekah Higgitt at the National Maritime Museum; Alan Perkins and staff at the Cambridge University Library; staff at the Digitale Bibliothek, Staats- und Universitätsbibliothek, Dresden; Isolde Müller at the Sternwarte Wien for her generous help with illustrations; and, of course, the Wellcome Trust and Anna Smith for their help and their fabulous image library.

I'm indebted to the following archives and libraries for their permission to quote from their manuscripts: Massachusetts Historical Society, Science and Technology Facilities Council and the Syndics of Cambridge University Library, American Philosophical Society, and the Royal Society, London.

I'm grateful for the unwavering support from Conville & Walsh, but this time in particular I'm indebted to Jake Smith-Bosanquet and Alexandra McNicoll for making this an international collaboration. And then, of course, lovely Patrick Walsh who is the most steadfast of friends and agents in the solar system.

Thank you to everybody at William Heinemann – although Drummond left us! Thank you to Jason Arthur, Laurie Ip Fung Chun for always being at the other end of the telephone and to Tom Avery for coming in at the last minute and working so unbelievably hard. I was lucky in the end. At Knopf, I would like to thank Edward Kastenmeier for his comments and support, Emily Giglierano and the wonderful Sara Eagle.

A huge thank you to Leo Hollis, who convinced me that I would be able to write this book despite the transit looming all too soon (and, of course, for reading the proposal); and another enormous thank you to Constanze von Unruh who is not only a great friend but who turned out to be a clever editor . . . again and again . . . thank you! I'm so grateful to Rebecca Carter, who rescued me once again when no one else seemed to care – you are truly the best; Olga and Tim for emergency Russian translations; Lisa O'Sullivan for opening her address book; and Tom Holland for Latin advice. Thank you to Beatrix Wulf for support, sea maps and other seafaring information. And to Julia-Niharika – I don't know where to begin – thank you for being my best friend, for our mad and delicious time in HH and for your smart comments on the chapters . . . and for so much more. My heart goes to Christian who was thrown into a crazy year of travels, moves, and deadlines but who kept me sane, fed me and supplied the best playlists at each ridiculous turn.

And as always, thank you to my lovely Linnéa for being such a fabulous (and clever) daughter – what would I do without you?

I could have not written this book without my parents Herbert and Brigitte Wulf – Herbert, thank you for reading all the chapters and for battling through (and translating) eighteenth-century Swedish and thank you, Brigitte, for translating French books and pages upon pages upon pages of eighteenth-century French journals and pamphlets on 'light-hearted' subjects such as longitude, navigation, astronomy and the solar parallax . . . what a feat, and much of it completed with a broken arm. You should blame my French teachers.

This book is dedicated to the wonderful Regan Ralph. Let's raise a glass of wine to international friendship. Thank you for being there!

Notes

Prologue

xxiii   Halley's essay: Halley, *Phil Trans*, 1714–16, vol. 29; Halley 1716 (Ferguson 1764)

xxiv   'young Astronomers': Halley 1716 (Ferguson 1764), p.14

xxv   Kant's '*Welteninseln*': Kant 1755

xxvi   'always been a principle': Winthrop 1769b, p.14

xxviii   Halley's diving bell: Halley, 'The Art of Living under Water: Or, a Discourse concerning the Means of Furnishing Air at the Bottom of the Sea, in Any Ordinary Depths', *Phil Trans*, 1714–16, vol. 29, p.10

xxviii   Halley's expeditions: Jardine 1999, p.24

xxviii   'talks, swears, and drinks': John Flamsteed, quoted in Ronan 1970, p.185

xxviii   Halley convincing Newton: Jardine 1999, p.34ff

xxviii   'his death-bed': Chappe 1770, p.81

xxviii   glass of wine: Armitage 1966, p.213

xxviii   'Indeed I could wish': Halley 1716 (Ferguson 1764), p.21

xxviii   'I recommend it therefore': *Ibid.*

xxix   London to Newcastle: Porter 2001, p.41

xxx   2,000 units of measure: Crump 2001, p.82

1 Call to Action

4   Delisle to Académie meeting: 30 April 1760, PV Académie 1760, f.257

4   'fit to walk': Benjamin Franklin to Polly Stevenson, 14 September 1767, BF online

4   'of all sorts': Duchess of Northumberland, May 1770, transcribed in Munck 2000, p.40

4   'Prince of Blood': *Ibid.*

4   'a prodigious Mixture': Benjamin Franklin to Polly Stevenson, 14 September 1767, BF online

4   'ugliest, beastly town': Horace Walpole to Thomas Gray, 19 November 1765, Lewis 1937–61, vol. 14, p.143

4   'what the heart': Louis-Sébastien Mercier, quoted in Jones 2006, vol. 10, point 5.1

5   Académie des Sciences: Hahn 1971, pp.21–22, 87–94

5   'membre de l'Académie': *Ibid.*, p.35

5   Halley's challenge: Halley, *Phil Trans*, 1714–16, vol. 29, pp.454–464; 30 April 1760, PV Académie 1760, f.257

6   Delisle biographical information: Woolf 1959, p.23ff; Nevskaia 1973, p.291ff

7   Delisle meeting Halley: Woolf 1959, p.30

7   Delisle 'greedy': Ulrik Scheffer to Wargentin, 23 September 1754, quoted in Lindroth 1967, p.397

7   'pester all and': La Caille to Wargentin, 1 Dec 1754, quoted in Widmalm 1992, p.49

7   'devouring gulf': Lalande to Wargentin, 4 March 1759, quoted in *Ibid.*

7   Delisle's conclusions: 21 November 1759, PV Académie 1759, f.770ff

7   Halley's calculations: Halley 1716 (Ferguson 1764), pp.19–20; see also Woolf 1959, p.55

7   'I have found': 21 November 1759, PV Académie 1759, f.770

7   Delisle's new calculations: 21 November 1759, PV Académie 1759, f.770ff; see also 5 June 1760, JBRS, vol. 24, f.596

8   Delisle's mappemonde: Woolf 1959, p.57ff

8   Delisle trained as surveyor: Sheehan and Westfall 2004, p.136

8   Delisle's distribution list for the mappemonde: reprinted in Woolf 1959, pp.209–211

8   mappemonde in France: Woolf 1959, pp.57–58

10   'here forlorn, he': Croarken 2003, p.285

10   'books of voyages': David Kinnebrook Sr to David Kinnebrook Jr, 9 January 1796, quoted in Croarken 2003, p.289

10   'indefatigable hard working': John Pond, quoted in Croarken 2003, p.286

10   Halley and colonial destinations: Halley 1716 (Ferguson 1764), p.20

11   Delisle letter to The Hague: Delisle to Bevis, 18 May 1760 and Delisle to Dirk Klinkenberg, 18 May 1760, Zuidervaart and Van Gent 2004, p.5

11   Delisle's claim to have Dutch cooperation: Woolf 1959, p.68

11   'usefulness of astronomy': Dirk Klinkenberg to Delisle, 6 June 1760, Woolf 1959, p.68

11   Delisle's method: 21 November 1759, PV Académie, 1759, f.771ff; Delisle 1760a; Woolf 1959, p.33ff

11   Le Gentil's departure from Brest: Le Gentil to Lanux, 15 September 1760, Le Gentil 1779 and 1781, vol. 2, pp.694–717

11   'not very well-to-do': Cassini 1810 (English translation in Sawyer Hogg 1951, p.39)

12   Le Gentil biographical information: Cassini 1810 (English translation in Sawyer Hogg 1951, p.39); Woolf 1959, pp.50–52, 58–60

12   'vain' theological arguments: Cassini 1810 (English translation in Sawyer Hogg 1951, p.39)

12   observe the 'heavens': *Ibid.*

12   Le Gentil to Pondicherry: Le Gentil 1760, pp.132–142

12   Le Gentil's permission to leave: the secretary of the Russian Academy of Sciences received a letter from his French colleagues on 3 January 1760, stating that Le Gentil would travel to Pondicherry. Le Gentil's travels were announced on 19 January 1760 at the meeting of the French Academy. Morosow 1954, p.432; Woolf 1959, p.58; Le Gentil 1760, p.139

12   'always zealous': and following quote, Le Gentil 1760, p.139

12   'great eagerness': Anon., 'Du Passage de Vénus sur le Soleil, Annoncé pour l'année 1761', *Histoire & Mémoires*, 1757, p.89; 19 April 1760, PV Académie 1760, f.239; Woolf 1959, p.60; Morosow 1954, p.432; 26 May/6 June 1760, Protocols, vol. 2, p.451

12   Pingré at Académie: Anon., 'Du Passage de Vénus sur le Soleil, Annoncé pour l'année 1761' *Histoire & Mémoires*, 1757, p.93

12   'worthy' and following quote: *Ibid.*

13   meeting RS on 5 June: 5 June 1760, JBRS, vol. 24, f.593ff

13   roads and hackney coaches: de Saussure, 26 October 1726, Muyden 1902, pp.166–167

13   sedan chairs: de Saussure, 26 October 1726, Muyden 1902, p.168; Lichtenberg to Ernst Gottfried Baldinger, 10 January 1775, Mare and Quarrell 1938, p.64

13   'made entirely of glass': Lichtenberg to Ernst Gottfried Baldinger, 10 January 1775, Mare and Quarrell 1938, p.6

13   'one shop jostles': Carl Philip Moritz , 2 June 1782, Nettel 1965, p.33

13   shopping and streets in London: Lichtenberg to Ernst Gottfried Baldinger, 10 January 1775, Mare and Quarrell 1938, pp.63–64

13   'Watchman's hoarse Voice': William Franklin to Elizabeth Graeme, 9 December 1757, BF online

13   description of meeting room of Royal Society: Christlob Mylius, 24 September 1753, Mylius 1787, p.74

13   'improvement of naturall': Jardine 1999, p.83

14   'Mémoire presented by Mr de Lisle': Delisle to RS, read on 5 June 1760, JBRS, vol. 24, f.596

14   'proper places': 19 June 1760, JBRS, vol. 24, f.84

15   Council 'unanimously' chose: 26 June 1760, CMRS, vol. iv, f.224

15   'if it were not attended': 26 June 1760, CMRS, vol. iv, f.224

15   flurry of activity and following descriptions: 26 June 1760, CMRS, vol. iv, ff.223–224

15   'what assistance might': 26 June 1760, CMRS, vol. iv, f.224

15   meeting RS 3 July: 3 July 1760, CMRS, vol. iv, p.225ff; see also Memorandum of Conversation between RS and directors of East India Company, 2 July 1760, RS MM/10, f.105

16   information on climate in Bencoolen: Lennox to Royal Society, 28 June 1760, RS MM/10, f.104

16   'in their Power' and following description: Memorandum of Conversation between RS and directors of East India Company, 2 July 1760 RS MM/10, f.105

16   'which (most likely)': Memorandum of Conversation between RS and directors of East India Company, 2 July 1760 RS MM/10, f.105

16   instructions to employees: Instructions sent by the Directors of East India Company to their several Presidencies, for observing the Transit of Venus, May 1760, RS MM/10, f.106; and 3 July 1760, CMRS, vol. iv, p.227

16   purchase of instruments: 3 July 1760, CMRS, vol. iv, p.237

16   budget for expeditions: 3 July 1760, CMRS, vol. iv, p.228

17   'Dr Halley, his Majesty's': RS MM/10, f.108

17   'now sending proper': *Ibid.*

17   'general Expectation': Lord Macclesfield to Duke of Newcastle, 5 July 1760, CMRS, vol. iv, f.230

17   'had been graciously': 14 July 1760, RS MM/10, f.108
17   Maskelyne appointed: 14 July 1760, CMRS, vol. iv, f.243
17   Maskelyne's love for astronomy: Nevil Maskelyne's Autobiographical
      Notes, transcribed in Howse 1989, Appendix B, p.215; see also Howse
      1989, p.15ff
17   'no celestial Phenomonen': Delambre, Life and Works of Dr Maskelyne,
      1813, p.2, RGO/226]
17   theories were 'sublime': Nevil Maskelyne to Charles Mason, 9
      November 1769, RGO 4/184 Letter 13
17   Maskelyne biographical information: Howse 1989

2 The French Are First

19   'in *different Ships*': CMRS, vol. iv, f.329
19   Le Gentil's journey and following description: based on Le Gentil
      to Lanux, 15 September 1760, Le Gentil 1779 and 1781, vol. 2,
      pp.694–717
20   'to place us between': Le Gentil to Académie, July 1760, read on 14
      February 1761, PV Académie, 1761, f.34
20   Le Gentil and lunar eclipse: *Ibid.*, f.35
20   'The fog seemed': *Ibid.*
20   been a 'relief': Le Gentil to Lanux, 15 September 1760, Le Gentil 1779
      and 1781, vol. 2, p.695
20   'better at sea': Le Gentil to Académie, July 1760, read on 14 February
      1761, PV Académie, 1761, f.35
20   'without tiring': *Ibid.*
20   Le Gentil arrival Mauritius: Le Gentil to Lanux, 15 September 1760,
      Le Gentil 1779 and 1781, vol. 2, p.713
21   'nicest and happiest': Le Gentil to Académie, July 1760, read 14
      February 1761, PV Académie, 1761, 14 February 1761, f.34
21   'one passenger who': *Ibid.*
21   British attacked French in India: Le Gentil to Lanux, 15 September 1760,
      Le Gentil 1779 and 1781, vol. 2, p.713; Le Gentil to Académie, July 1760,
      read 14 February 1761, PV Académie, 1761, 14 February 1761, f.36
21   'destined to the siege': Le Gentil to Académie, July 1760, read 14
      February 1761, PV Académie, 1761, 14 February 1761, f.36
21   'conveying their artillery': *Ibid.*
21   hurricane destroying fleet: *Ibid.*, f.37
21   'I do not know when': *Ibid.*, f.36
22   'chimerical projects': Le Gentil to Lanux, 6 February 1761, Le Gentil
      1779 and 1781, vol. 2, p.719
22   Le Gentil considering different viewing places: Le Gentil (Ebeling)
      1781, pp.30–31
22   'always foggy and': Le Gentil to Duc de Chaulnes, 6 September 1761,
      read on 30 January 1762, PV Académie, 1762, f.20
22   Rodrigues's climate: Le Gentil to Lanux, 6 February 1761, Le Gentil
      1779 and 1781, vol. 2, pp.719–720
22   'I am without hope': *Ibid.*, p.720
22   'Life is horribly': Le Gentil to Académie, July 1760, read 14 February
      1761, PV Académie, 1761, 14 February 1761, f.36
22   attacks of dysentery: Le Gentil to Lanux, 6 February 1761, Le Gentil
      1779 and 1781, vol. 2, p.719; Le Gentil (Ebeling) 1781, p.29

22   'mortification and concern': Le Gentil to Lanux, 6 February 1761, Le Gentil 1779 and 1781, vol. 2, p.719

23   report on Pingré's destination: read on 20 August 1760, Chabert 1757b, pp.43–49; see also Woolf 1959, p.64ff

23   the Académie's 'zeal': de Chabert, read on 20 August 1760, *Histoire & Mémoires*, 1757, p.43

23   'those precious moments': Lalande, 'Mémoire sur le passages de Vénus devant le disque du Soleil, en 1761 et 1769', *Histoire & Mémoires*, 1757, p.250

23   century had 'envied': *Ibid.*

23   'future' would blame: *Ibid.*

23   Portuguese or Dutch ports: de Chabert, read on 20 August 1760, *Histoire & Mémoires*, 1757, pp.44–45

23   'dangerous for foreigners': *Ibid.*, p.47

23   'to be supplanted': *Ibid.*

23   'not alarmed by': Anon., 'Du passage de Vénus sur le Soleil, Annoncé pour l'année 1761', *Histoire & Mémoires*, 1757, p.84

23   Pingré and gout: 10 January 1761, Pingré 2004, p.44

23   Pingré biographical information: Woolf 1959, pp.97–101; Armitage 1953, p.48ff

24   'no doubt surpass': de Chabert, 20 August 1760, *Histoire & Mémoires*, 1757, p.46

24   write to Holland and Portugal: Anon., 'Du Passage de Vénus sur le Soleil, Annoncé pour l'année 1761', *Histoire & Mémoires*, 1757, p.90

24   'many obstacles': *Ibid.*

24   support from local administration: *Ibid.*, p.92

24   Pingré's farewell dinner: Woolf 1959, p.101; Pingré 2004, p.39

24   Pingré and food: see his many entries in his journal in Pingré 2004

24   'extremely flattered': Pingré 2004, p.39

24   'the first to be frightened' and following quotes: *Ibid.*

25   'my liberty, my health': *Ibid.*

25   coach to Lorient: Woolf 1959, p.101

25   Pingré's wait in Lorient: Pingré 2004, p.39ff; see also Woolf 1959, pp.101–102

25   thirty-eight cannons: 10 January 1761, Pingré 2004, p.44

25   Pingré's departure: 9 January 1761, *Ibid.*, p.43

25   Pingré and galley supplies: 9 January 1761, *Ibid.*

25   'paid their tribute': 10 January 1761, *Ibid.*, p.44

26   *'not to molest'*: Pingré's passport, 25 November 1760: Appendix III D, Woolf 1959, p.208

26   'to proceed without': *Ibid.*

26   description of 10 January 1761 and enemy vessels: 10 January 1761, Pingré 2004, pp.44–45

26   'without firing a shot': 10 January 1761, *Ibid.*, p.45

26   'grand ballroom': 15 January 1761, *Ibid.*, p.49

26   monotonous ennui: Pingré's journal entries for this time, in Pingré 2004

26   'alone than in the': 15 March 1761, Pingré 2004, p.91

26   boring life on board: 15 March 1761, *Ibid.*

27   imprisoned in the Bastille: 15 March 1761, *Ibid.*

27   'an excellent vessel': 30 January 1761, *Ibid.*, p.61

27   flying fish: 27 January 1761, *Ibid.*, p.56
27   sea 'on fire': 22 January 1761, *Ibid.*, p.54; and many other times
27   sailor falls: 2 February 1761, *Ibid.*, p.62
27   'equator baptism': 14 February 1761, *Ibid.*, p.74
27   'le père *la Ligne*': 14 February 1761, *Ibid.*, p.74
27   Pingré and southern sky: Pingré, 24 February 1761, *Ibid.*, p.81
27   'liquor gives us': Pingré MS journal, quoted in French in Woolf 1959, p.106
27   'not with the bottle': *Ibid.*
27   Meeting the *Le Lys*: Pingré, 8 April 1761 (and following days), Pingré 2004, pp.105-122
27   reduced speed: 19 and 20 April 1761, *Ibid.*, pp.117-118
28   'completely useless': 12 April 1761, *Ibid.*, p.110
28   Pingré heard Le Gentil was in Mauritius: 12 April 1761, Pingré 2004, p.110
28   disputes with the *Le Lys*: 10, 12 April 1761, Pingré 2004, pp.107, 109-110; Pingré to Marion and Marion to Pingré, 13 April 1761, *Ibid.*, pp.112-114
28   'holiest of his duties': Pingré to Marion, 13 April 1761, *Ibid.*, p.112
28   'whole of Europe': Pingré to Marion, 13 April 1761, *Ibid.*, p.112
28   on course to Rodrigues: 1 May 1761, *Ibid.*, p.122
28   fresh fruit and meat: Pingré to Marion, 12 April 1761, *Ibid.*, p.109
28   'throw him over': Pingré MS journal, quoted in French, in Woolf 1959, p.108
28   Rodrigues on the horizon: 3 May 1761, Pingré 2004, p.124
28   one last attempt: Pingré, 2 May 1761, *Ibid.*, pp.122-123
28   Le Gentil's passage to Pondicherry: Le Gentil to Lanux, 16 July 1761, Le Gentil 1779 and 1781, vol. 2, p.721; Le Gentil (Ebeling) 1781, p.31
29   find another viewpoint: Le Gentil to Lanux, 16 July 1761, Le Gentil 1779 and 1781, vol. 2, p.721
29   description of sea journey from Mauritius: Le Gentil to Lanux, 16 July 1761, Le Gentil 1779 and 1781, vol. 2, p.721ff, Le Gentil (Ebeling) 1781, p.31ff
29   island of Socotra: Le Gentil to Lanux, 16 July 1761, Le Gentil 1779 and 1781, vol. 2, p.724; Le Gentil (Ebeling) 1781, p.32
29   'golden sequins': Le Gentil to Lanux, 16 July 1761, Le Gentil 1779 and 1781, vol. 2, p.728
29   'golden columns': *Ibid.*
30   the south-west monsoon: *Ibid.*, p.736
30   Le Gentil saw lights: *Ibid.*, p.742
30   news about Mahé and Pondicherry: *Ibid.*, p.743
30   'my great vexation': Le Gentil (Ebeling) 1781, p.33
30   storm: Le Gentil to Lanux, 16 July 1761, Le Gentil 1779 and 1781, vol. 2, p.744

3 Britain Enters the Race

31   Liquor allowance: of the £800 allocated for the trip, almost £200 was set aside for 'liquor'. Estimate St Helena, 21 July 1760, CMRS, vol. iv, ff.246; 5 August 1760, CMRS, vol. iv, ff.251-253; 5 August 1760, RS MM/10, f.111-112; see also Howse 1989, p.25

31   'intimately connected with': 21 July 1760, CMRS, vol. iv, f.254

31   mapping of Russia: Vucinich 1984, p.22

32   Pingré's ship across two islands: 30 January 1761, Pingré 2004, p.61

32   Longitude and longitude calculations: Richardson 2005, pp.34–35; Howse 1989, pp.92–93

33   *Gulliver's Travels* and longitude: Swift 1826, p.83

33   'ordered a Ship': Admiralty to Royal Society, 30 July 1760, read on 5 August 1760, CMRS, vol. iv, f.250

33   Maskelyne and longitude: Howse 1989, pp.30, 42

34   British exports colonies: Longford 1992, pp.168–169

34   French used same argument: Chabert, *Histoire & Mémoires*, 1757b, p.44, read on 20 August 1760

34   French colonial trade: McClellan and Regourd 2000, p.49

34   Maskelyne's departure: Maskelyne, 'Journal of Voyage to St Helena, 1761–1762', RGO 4/150; Nevil Maskelyne to William Watson, 17 January 1761, read at the RS on 22 January 1761, JBRS, vol. 25, f.18ff

34   Pingré and Madeira: 20 January 1761, Pingré 2004, p.53

34   convoy accompanying Maskelyne's ship: Howse 1989, pp.28–29

35   Maskelyne's lunar method: Howse 1989, p.29–30, Nevil Maskelyne's Autobiographical Notes, transcribed in Howse 1989, Appendix B, p.218; Maskelyne, 'Journal of Voyage to St Helena, 1761–1762', RGO 4/150

35   'of the Motions': Jardine 1999, p.14

35   Mayer's lunar tables: Howse 1989, p.30; Nevil Maskelyne's Autobiographical Notes, transcribed in Howse 1989, Appendix B, p.218

35   Maskelyne's observation notes: Maskelyne, 'Journal of Voyage to St Helena, 1761–1762', RGO 4/150

35   Maskelyne and Mayer's lunar tables: Howse 1989, pp.29–30; Nevil Maskelyne's Autobiographical Notes, transcribed in Howse 1989, Appendix B, p.218

35   'My principal attention': Maskelyne to Birch, 13 May 1761, quoted in Howse 1989, p.30

36   'ascertain his Longitude': Maskelyne to Birch, 13 May 1761, quoted in *Ibid*.

36   Maskelyne's liquor supplies: wine merchant bill, 23 December 1760, RS MM/10, f.151

36   'a very agreeable voyage': Maskelyne to Birch, 13 May 1761, quoted in Howse 1989, p.30

36   St Helena appeared: 5 April 1761, Maskelyne, 'Journal of Voyage to St Helena, 1761–1762', RGO 4/150

36   weather on arrival: *Ibid*.

37   'almost to overhang': Joseph Banks's Account of St Helena, Beaglehole 1962, p.478

37   'came to an Anchor': 6 April 1761, Maskelyne, 'Journal of Voyage to St Helena, 1761–1762', RGO 4/150

37   having to find observatory: Maskelyne to Birch, 13 May 1761, quoted in Howse 1989, p.30

37   RS appoints Mason and Dixon: 11 and 25 September, 10 October

1760, CMRS, vol. iv, ff.253–254, 256; 11 and 19 September, RS MM/10, f.108

37 Mason biographical information: Cope 1951, p.232; Robinson 1949, p.134

37 Dixon biographical information: Cope and Robinson 1951, p.56; Robinson 1950, pp.272–274

37 'nothing can exceed': Thomas Evans, 1796–98, quoted in Croarken 2003, p.285

37 'drinking to excess': Dixon was expelled on 28 October 1760, Robinson 1950, p.273

37 Mason and Dixon salaries: 11 September 1760, CMRS, vol. iv, f.253; for their contracts 23 October 1760, CMRS, vol. iv, f.265; Mason's salary at the Royal Observatory: Croarken 2003, p.293

38 captain interviewed: 16 October 1760, CMRS, vol. iv, f.262

38 'every thing in their power': 16 October 1760, RS MM/10, f.114

38 'Companys Expence': Ibid.

38 Mason and Dixon to Portsmouth: 17 November 1760, CMRS, vol. iv, f.269; Dixon attended this meeting and reported that he would leave London the next day.

38 Delayed by contrary winds: Mason to Royal Society, 8 and 19 December 1760, RS MM/10, ff.124, 126

38 Battle HMS Seahorse: Charles Mason to Morton, 12 January 1761, RS MM/10, f.128; Captain Smith, 11 January 1761, in Edinburgh Magazine, March 1761, p.161; Annual Register 1761, p.54

38 'crouding down upon': Captain Smith, 11 January 1761, in Edinburgh Magazine, March 1761, p.161

38 'pistol-shot': Ibid.

39 'I believe mortal': Charles Mason to Charles Morton, 12 January 1761, RS MM/10, f.128

39 Mason and Dixon's instruments: Mason and Dixon, Phil Trans, 1761–62, p.394

39 broken stands: Receipt for Returning the Instruments, 18 May 1762, RS MM/10, f.146

39 'shattered' ship: Charles Mason to Charles Morton, 12 January 1761, RS MM/10, f.128; and 21 January 1761, CMRS, vol. iv, f.285

39 'absolutely impossible': Charles Mason and Jeremiah Dixon to Charles Morton, 25 January 1761, RS MM/10, f.130

39 letters Mason and Dixon: Charles Mason to Richard Bradley, 25 January 1761; Mason and Dixon to Charles Morton, 25 January 1761; Charles Mason and Jeremiah Dixon to Thomas Birch, 27 January 1761, RS MM/10, ff.129–131

39 letters read at RS: 31 January 1761, CMRS, vol. iv, ff.288–290

40 viewing transit in Scanderoon: Charles Mason and Jeremiah Dixon to Charles Morton, 25 January 1761, RS MM/10, f.130

40 would 'obey' commands and following quotes: Ibid.

40 'no reason why I' and following quotes: Charles Mason to Richard Bradley, 25 January 1761, RS MM/10, f.129

40 'to go on Board': 31 January 1761, CMRS, vol. iv, f.290

40 RS 'surprised' and following quotes: Royal Society to Charles Mason and Jeremiah Dixon, 31 January 1761, RS MM/10, f.132

40   'absolutely refused to proceed': Captain Smith to Royal Society, 31 January 1761, RS MM/10, f.133

41   Morton to Admiralty: 22 January and 5 February 1761, JBRS, vol. 25, ff.16, 33–34

41   'preparations now making': Benedict Ferner to Thomas Birch, 13 January 1761, read at the RS on 29 January 1761, JBRS, vol. 25, f.25

41   'peremptory refusal': 5 February 1761, *Ibid.*, f.34

41   'with the utmost Severity': Royal Society to Charles Mason and Jeremiah Dixon, 31 January 1761, RS MM/10, f.132

41   'We are sorry' and following quote: Charles Mason and Jeremiah Dixon to RS, 3 February 1761, RS MM/10, f.134

4 To Siberia

42   Academy and Peter the Great: Home 1973, p.75; Schulze 1985, p.305ff

42   salaries for foreigners: Schulze 1985, p.310

42   'brazenly derided': Leonard Euler to Mikhail Lomonosov, 11 January 1755, Juskevic and Winter 1959–76, vol. 3, p.202

42   had 'commended': 26 May/6 June 1760, Protocols, vol. 2, p.451; the letter from France arrived in January 1760, Morosow 1954, p.432. The Russians still used the Julian Calendar which was ten days 'earlier' than the Gregorian Calendar which was used in Europe and America. When referring to Russian sources, I have always listed both dates – in this case the 26 May 1760 is the date used in Russia and 6 June 1760 is the date according to the much more widely used Gregorian Calendar.

42   Russian academicians wrote to Delisle: 19 April 1760, PV Académie 1760, in Woolf 1759, p.60; Morosow 1954, p.432; 26 May/6 June 1760, Protocols, vol. 2, p.451

43   Chappe recruited: Anon., 'Du Passage de Vénus sur le Soleil, Annoncé pour l'année 1761', *Histoire & Mémoires*, 1757, p.89

43   Chappe biographical information: Fouchy 1769, p.163ff; Armitage 1954, p.278ff; Woolf 1959, p.115ff

43   aristocratic clients: Fouchy 1769, pp.163–164

43   'He liked fame': *Ibid.*, p.172

43   Chappe's journey to Tobolsk: unless otherwise referenced, Chappe 1770

43   accidents after departure: Chappe 1770, p.2

44   'only for a few hours': *Ibid.*

44   'account of a quarrel': *Ibid.*, p.4

44   'take the veil': *Ibid.*, p.5

44   'in continual apprehensions': *Ibid.*, p.8

44   accident between Brno and Nový Jičín, and following description: *Ibid.*

44   Chappe's character: Fouchy 1769, pp.171–172

45   'I began to fear': Chappe 1770, p.8

45   'not before experienced': *Ibid.*, p.9

45   'eleven degrees below o': *Ibid.*, p.15

45   carriage accidents: *Ibid.*, pp.17–18

45   'top to bottom': *Ibid.*, p.18

45   walking in mountains: *Ibid.*, p.19

45   clouds of snow: *Ibid.*, p.24

45   'could not stand it': *Ibid.*, p.21

45   stuck once again: *Ibid.*, p.22

45   interpreter 'in liquor': *Ibid.*, p.23

45   'continual delays': *Ibid.*, p.20

45   measuring petticoats: *Ibid.*, p.9

45   'strictly virtuous': *Ibid.*, p.10

45   'slenderness of their': *Ibid.*, p.11

45   'well-shaped servant': *Ibid.*, p.21

46   arrival St Petersburg: *Ibid.*, p.25

46   Chappes's Academy meeting: 9/20 February 1761, Protocols, vol. 2, p.463

46   row between Aepinus and Lomonosov: Home 1973, p.81ff; Pekarsky 1870–73, vol. 2, pp.698–733

46   Aepinus's career: Home 1973, p.89

46   Lomonosov's temperament: Stepan Rumovsky to J. A. Euler, 2 December 1764, *Ibid.*, p.77

46   Lomonosov biographical information: Menshutkin 1952; Vucinich 1984, p.15ff; Huntington 1959, p.295ff; Home 1973, pp.76–77

47   'any service to the': Pekarsky 1870–73, vol. 2, p.698

47   'dared not utter': Pushkin, quoted in Menshutkin 1952, p.185

47   drunken brawl: *Ibid.*, p.39

47   indecent pranks: *Ibid.*

47   stupid 'lapdogs': Lomonosov to Leonard Euler, spring 1765, Juskevic and Winter 1959–1976, vol. 3, p.202

47   Aepinus's essay: Pekarsky 1870–73, vol. 2, p.701; Home 1973, p.80ff

47   Lomonosov's criticism of the essay: Pekarsky 1870–73, vol. 2, p.701

47   Lomonosov undermining Aepinus: *Ibid.*, p.700ff

47   'false rumours throughout': 19 December 1760, Protocols, vol. 2, p.459

47   Aepinus's reaction to Lomonosov: Pekarsky 1870–73, vol. 2, p.702; 1/12 and 8/19 December 1760 Protocols, vol. 2, pp.458–459

47   scheme was 'flawed' and following quotes: 8/19 December 1760, Protocols, vol. 2, pp.459–460

48   Russian expeditions: 13/24 November 1760, *Ibid.*, p.458

48   'honour' of the empire: Morosow 1954, p.432

48   preparations for expeditions: 24/30 November 1760, Protocols, vol. 2, p.458; Morosow 1954, p.432

48   Russian expedition leaves in mid-January 1761: Benedict Ferner to Thomas Birch, 13 January 1761, read at the RS on 25 January 1761, JBRS, vol. 25, f.25; 24/30 November 1760, Protocols, vol. 2, p.458; Morosow 1954, p.432

48   Chappe should view transit closer to St Petersburg: Chappe 1770, pp.25–26

48   'so much advantage': *Ibid.*, p.25

48   'easily comprehended': *Ibid.*, p.26

48   'lover and protector': *Ibid.*

49   departure from St Petersburg and following description: *Ibid.*, p.26ff

49   'mend my clocks': *Ibid.*, p.26

49   race against thaw: *Ibid.*

50   'greatest velocity': *Ibid.*, p.25

50   'smooth as glass': *Ibid.*, p.38

50   'inconceivable swiftness': *Ibid.*
50   standing 'upright': *Ibid.*
50   sledges constantly overturned: *Ibid.*, p.63
50   horse fell into river: *Ibid.*, p.31
50   liable to be swallowed: *Ibid.*, p.61
50   Chappe on Russian women: *Ibid.*, pp.65, 72
51   ice melting: *Ibid.*
51   'overtaken by a thaw': *Ibid.*, p.71
51   refused to cross river: *Ibid.*, p.73
51   'attended with an universal': *Ibid.*, p.74
51   copious amounts of brandy: *Ibid.*, p.75
51   hurried his drunken men: *Ibid.*, p.76

5 Getting Ready For Venus

52   observatory in Hofgarten: Johann Georg von Lori to Johann Heinrich Samuel Formey, 15 July 1761, Spindler 1959, p.429; Westenrieder 1784, p.74
52   instruments on garden wall: Prosper Goldhofer to Johann Georg von Lori, 28 May 1761, Spindler 1959, p.408
52   library in Warsaw: Stefan Luskina at the Zaluski Public Library in Warsaw, *Das Neuste aus der Anmuthigen Gelehrsamkeit*, 1761, vol. 1, p.501
52   information about Academy of Sciences and politics: Frängsmyr 1989, p.2ff; Lindroth 1952, pp.16–19
54   'useful' science: Frängsmyr 1989, p.2
54   'silkworm farming': *Ibid.*, p.5
54   Wargentin biographical information: Frängsmyr 1989, pp.7–8, 50–56; Lindroth 1952, pp.105–110
54   Wargentin's love of astronomy: Lindroth 1952, pp.105–106
54   'a kind of brotherhood': Wargentin, quoted in Frängsmyr 1989, p.7
54   'Before my time': Wargentin, quoted in McClellan 1985, p.171
54   exchange of journals: *Ibid.*
55   Wargentin encourages astronomers: Lindroth 1967, p.401
55   Delisle sent mappemonde: *Ibid.*, p.403
55   Scandinavia an important counterpart: Wargentin, KVA Abhandlungen 1761b, p.142
55   Wargentin ordered instruments: 30 April 1760, KVA Protocols, p.618
55   organised Lapland expedition: Nordenmark 1939, pp.175–176; Lindroth 1967, p.401; 21 January 1761, KVA Protocols, p.634
55   Linnaeus and Lapland: Sörlin 2000, p.57
55   Planman expedition: Anders Planman to Wargentin, 18 November 1760, MS Wargentin, KVA, Center for History of Science; Lindroth 1967, p.401; Nordemark 1939, p.175
55   'thinnest' lenses: Planman to Wargentin, 18 November 1760, MS Wargentin, KVA, Center for History of Science
55   Planman's departure: Planmann left Uppsala after drawing the funds for his expedition, KVA, Verification no. 116, 12 January, 4 and 12 February; 21 January 1761, KVA Protocols
56   severe winter: Wargentin, KVA Abhandlungen, 1761, p.143
56   'superb stalactites': Acerbi 1802, vol. 1, p.184

56   'raised perpendicularly': *Ibid.*, p.185

56   'a wolf or bear rolling': *Ibid.*

56   Planman's illness: Planman to Wargentin, 16 April 1761, MS Wargentin, KVA, Center for History of Science

56   'dreary silence': Acerbi 1802, vol. 1, p.228

56   Planman walking: Planman to Wargentin, 16 April 1761, MS Wargentin, KVA, Center for History of Science

56   Planman's journey: *Ibid.*

56   Planman's arrival in Kajana: *Ibid.*

57   Winthrop's expedition: Winthrop 1761, p.7

57   Winthrop's article: 2 April 1761, *Boston News-Letter*; 6 April 1761, *Boston Post Boy*; 10 April 1761, *New Hampshire Gazette*

57   'may ultimately be': Extracts of the Votes of the Hon. House of Representatives, Winthrop 1761, p.22

57   'do Credit to': *Ibid.*, p.23

57   'shall judge proper': *Ibid.*, p.24

57   'carrying on an illicite': *Edinburgh Magazine,* April 1761, p.187

57   expeditions to Bencoolen and St Helena: 13 October 1760, *Boston Evening Post*

57   Wargentin's instructions for amateurs: 28 May 1761, *Inrikes Tidningar,* 1761, vol. 47

57   viewing transit with smoked glasses: *Ibid.*

58   Delisle's pamphlet: 2 May 1761, PV Académie, 1761, f.85

58   whimsical 'dialogue': Martin 1759

58   'for those who do': Ferguson 1761, p.2

58   Ferguson's book at RS: 19 February 1761, JBRS, vol. 25, f.52

58   Ferguson and filling seats twice: Millburn 1976, p.85

58   Martin's advertisements: 28 May and 1, 3 June 1761, *Public Advertiser*

58   RS informing Delisle about Mason and Dixon: Woolf 1959, p.91

58   chaplain of the British community to RS: 12 March 1761, JBRS, vol. 25, ff.71–73

58   search for employee of Dutch East India Company: Dirk Klinkenberg to Delisle, 29 November 1760, Zuidervaart and Van Gent 2004, p.6

60   Chappe informing Russian Academy: 9/20 February 1761, Protocols, vol. 2, p.463

60   news of Pingré in Germany: Prosper Goldhofer to Johann Georg von Lori, 29 March 1761, Spindler 1959, p.390

60   information about Chappe to Berlin: 13/24 March 1761, Gerhardt Friedrich Müller to Leonard Euler, Juskevic and Winter 1959–76, vol. 1, p.172

60   information about Chappe to Leipzig: Gottfried Heinsius to Leonard Euler, 4 April 1761, Juskevic and Winter 1959–76, vol. 3, p.121

60   'to go to Batavia': Le Gentil to French Academy of Science, July 1760, read on 14 February 1761, PV Académie, 1761, f.37

60   news that Chappe had left St Petersburg: 20 May 1761, PV Académie 1761, f.91

60   'an universal tremor': Chappe 1770, p.76

60   severe spring floods: *Ibid.*, p.77

61   Maskelyne's search for observatory: Maskelyne to Birch, 13 May 1761, quoted in Howse 1989, pp.30–31

61   'inevitably be precipitated to': Kindersley 1777, pp.294–295

61 'almost perpetually covered': Maskelyne to Birch, 13 May 1761, quoted in Howse 1989, p.31

62 Halley on St Helena: Cook 1998, p.73ff

62 'infested' with clouds: Nevil Maskelyne's Autobiographical Notes, transcribed in Howse 1989, Appendix B, p.217

62 'below Halley's Mount' and following quotes: Maskelyne to Birch, 13 May 1761, quoted in Howse 1989, p.31

62 Maskelyne's astronomical preparations: Maskelyne, 'Journal of a Voyage from England to St Helena', RGO 4/150; Maskelyne, 'Observations at St Helena', RGO 4/2

62 'thro' the tardiness': 10 January 1765, JBRS, vol. 26, p.192

62 'utmost importance': Maskelyne to William Watson, 17 January 1761, Maskelyne, *Phil Trans*, 1761-62a, vol. 52, p.27

62 Mason and Dixon at Cape: Mason and Dixon, *Phil Trans*, 1761-62, vol. 52, p.380

62 French had taken Bencoolen: Charles Mason to RS, 6 May 1761, RS MM/10, f.135

63 instructions from Astronomer Royal: Bradley 1832, p.388

63 letter to RS: Jeremiah Dixon to Thomas Birch, 6 May 1761, original at BL, copy at APS

63 'distant part of the Globe': Joseph Banks's Account of Cape Town, Beaglehole 1962, p.449

63 description of Cape Town: *Ibid.*, p.456; Kindersley 1777, pp.56-57

64 oozing Dutch neatness: *Ibid.*, p.449ff; Kindersley 1777, p.53

64 'Foreigners . . . find themselves': Kindersley 1777, p.65

64 'common method of': Joseph Banks's Account of Cape Town, Beaglehole 1962, p.457

64 Mason and Dixon's lodgings: in their expenses Mason and Dixon list a 'Mr Zeeman' as their landlord. 'Expenses for the Royal Society', original at BL, copy at APS

64 botanical garden: Joseph Banks's Account of Cape Town, Beaglehole 1962, pp.458-460; Kindersley 1777, p.53

64 hired a carriage: 'Expenses for the Royal Society', original at BL, copy at APS

64 'Dutch are so slow' and following quote: Charles Mason to RS, 6 May 1761, RS MM/10, f.135

64 help to complete observatory: Jeremiah Dixon to Thomas Birch, 6 May 1761, original at BL, copy at APS

64 canvas and observatory: Mason and Dixon list '14 yards of canvas' in their expenses, 'Expenses for the Royal Society', original at BL, copy at APS

64 'any part of the heavens': Mason and Dixon, *Phil Trans*, 1761-62, vol. 52, p.379

64 'near all the time': *Ibid.*, p.382

65 locals believed Chappe was a magician: Chappe 1770, p.79

65 'should attempt to pull': *Ibid.*, p.80

65 Chappe and lunar eclipse: *Ibid.*, p.78

65 Planman's tower of armchairs: Anders Planman to Wargentin, 21 May 1761, MS Wargentin, KVA, Center for History of Science

66 Mason and Dixon and lunar eclipse: Mason and Dixon, *Phil Trans*, 1761-62, vol. 52, p.380

66   Pingré and lunar eclipse: 19 May 1761, Pingré 2004, p.133
66   Pingré's arrival Mauritius: 7 May 1761, *Ibid.*, pp.127–128
66   governor reassured Pingré: 7 May 1761, *Ibid.*, p.128
66   only eight days to Rodrigues: 14 April 1761, *Ibid.*, p.114
66   Pingré's journey to Rodrigues: 8–28 May 1761, *Ibid.*, pp.129–138
66   'that filled me with': 26 May 1761, *Ibid.*, p.137
66   'The calm continued': Pingré MS journal, quoted in Woolf 1959, p.109
66   'desired island': 28 May 1761, Pingré 2004, p.137
66   turtles on Rodrigues: *Ibid.*, Appendix 'Voyage à l'île de Rodrigue', pp.367–368
66   governor's residence Rodrigues: 28 May 1761, Pingré 2004, p.185
67   'We had no time': Pingré 2004, Appendix 'Voyage à l'île de Rodrigue', p.371
67   location of observatory: *Ibid.*, p.366
67   description of temporary observatory: Pingré, *Histoire & Memoires*, 1761b, p.414
67   'eaten by rust': Pingrée MS journal, quoted in Woolf 1959, p.110
67   rats chewing a pendulum: Pingré 2004, Appendix 'Voyage à l'île de Rodrigue', p.368

6 Day of Transit, 6 June 1761

68   Delisle's eyesight: Nevskaia 1973, p.309
68   Delisle's inspiring character: *Ibid.*, p.291
71   Le Gentil's geographical position: Le Gentil to Lanux, 16 July 1761, Le Gentil 1779 and 1781, vol. 2, p.748
71   Le Gentil's transit day: description based on *Ibid.* pp.746–751
71   not to be 'idle' and following quote: *Ibid.*, p.747
71   telescope fixed to timber beam: *Ibid.*
72   difficulties in focusing: *Ibid.*
72   Le Gentil saw Venus: *Ibid.*, p.748
72   'edge of the Sun': *Ibid.*, p.749
72   Venus 'exited': *Ibid.*
72   'far from' required precision: *Ibid.*, p.750
72   Chappe's previous evening and following description: Chappe 1770, pp.80–82
72   'perfect stillness': *Ibid.*, pp.80–81
72   'state of despondency' and following quotes: *Ibid.*, p.81
73   Chappe woke assistants: *Ibid.*, p.82
73   'dreadful agitations': *Ibid.*
73   red shadow of sun: *Ibid.*
73   'new kind of life' and following quotes: *Ibid.*
73   'shared my happiness': *Ibid.*
73   tent for visitors: *Ibid.*, p.80
73   doubled his guards: *Ibid.*, p.83
73   'sun a thousand times' and following quotes: *Ibid.*
74   rain on Rodrigues: Pingré 2004, Appendix 'Voyage à l'île de Rodrigue', p.369
74   Pingré's day of transit: and following descriptions, *Ibid.*, p.371ff; Pingré, *Phil Trans*, 1761–62, vol. 52, pp.371–377
74   Pingré's 'observatory': Pingré, *Histoire & Memoires*, 1761b, p.414

74   Pingré couldn't see anything: Pingré, *Phil Trans*, 1761–62, vol. 52, p.371

74   other observers on Rodrigues: June 1761, Pingré 2004, p.186

74   'cannot be depended upon': Pingré, *Phil Trans*, 1761–62, vol. 52, p.376

74   Pingré short-sighted: *Ibid.*, p.375

74   'disordered my instrument': *Ibid.*, p.376

74   cloud in front of the sun: Pingré 2004, Appendix 'Voyage à l'île de Rodrigue', p.371

74   'faint' shape: Pingré, *Phil Trans*, 1761–62, vol. 52, p.374

75   'certainly ended': *Ibid.*

75   'the astronomers of all': Pingré, MS journal, quoted in Woolf, p.111

75   row between Aepinus and Lomonosov: Pekarsky 1870–73, vol. 2, p.730ff; Home 1973, pp.88–89; Morosow 1954, p.433ff

75   'no knowledge on': Pekarsky 1870–73, vol. 2, p.732

75   Aepinus's response to Lomonosov: *Ibid.*, pp.732–733

76   too much noise: *Ibid.*, pp.731–732

76   Lomonosov's letter: *Ibid.*, pp.731–733

76   co-observers during astronomical events: *Ibid.*, p.731

76   'catechism in school': Lomonosov quoted in Morosow 1954, p.443

76   'count seconds without': Pekarsky 1870–73, vol. 2, p.732

76   'simply madness': *Ibid.*, p.733

76   Aepinus not observing transit: Morosow 1954, p.444

76   Lomonosov's day of transit and following descriptions: Meadows 1966, pp.118–119

76   'seemed to be disturbed': Lomonosov, quoted in *Ibid.*, p.118

76   rest 'weary eye': Lomonosov, quoted in Menshutkin 1952, p.147

76   'a little pimple': Lomonosov, quoted in Meadows 1966, p.118

77   'refraction of the solar': Lomonosov, quoted in *Ibid.*, p.119

77   life on Venus: Meadows 1966, p.120; Menshutkin 1952, p.148

77   Planman's day of transit and following descriptions: Planman to Wargentin, 11 June 1761, MS Wargentin, KVA, Center for History of Science; Wargentin, KVA Abhandlungen, 1761, p.143; for Planman's expedition: Nordenmark 1939, p.176; Lindroth 1967, p.401

77   Planman's observatory: Lindroth 1967, p.401

77   thick smoke and fire and following descriptions: Planman to Wargentin, 11 June 1761, MS Wargentin, KVA, Center for History of Science; Wargentin, KVA Abhandlungen, 1761 p.143

77   ordered a 'ban': Planman to Wargentin, 11 June 1761, MS Wargentin, KVA, Center for History of Science

78   'nuptials of Venus': *Ibid.*

78   Wargentin's day of transit Stockholm and following descriptions: Wargentin, KVA Abhandlungen, 1761, p.151ff; Lindroth 1967, p.402ff; Nordenmark 1939, p.176ff; Lindroth 1967, p.402

78   other observers in Stockholm: Johan Carl Wilcke, Samuel Klingenstierna, Johan Gabriel von Seth, Pehr Lehnberg, Carl Lehnberg and Jacob Gadolin, Wargentin, KVA Abhandlungen, 1761, p.152

78   calling out time: *Ibid.*

78   to be 'boiling': *Ibid.*

78   Wargentin unsure about time: *Ibid.*, p.153

79   glowing halo: *Ibid.*

79   'greatest diligence': *Ibid.*, p.154
79   'shooting': *Ibid.*
79   'narrow ring': *Ibid.*, p.155
79   range of times: *Ibid.*, p.156
79   'not agree so near': Wargentin to John Ellicot, 7 August 1761,
       Wargentin, *Phil Trans*, 1761–62, vol. 52, p.215
79   Mason and Dixon's day of transit Cape: Mason and Dixon, *Phil Trans*,
       1761–62, vol. 52, pp.383–384
79   previous six weeks at Cape: *Ibid.*, p.382
79   'a thick haze' and following description: *Ibid.*, p.383
80   'hid by a cloud': *Ibid.*, p.384
80   Maskelyne prepared his instruments: Maskelyne, 'Observations at St
       Helena', RGO 4/2
80   clock had stopped: 2 June 1761, *Ibid.*
80   Maskelyne's day of transit and following descriptions: Maskelyne,
       Phil Trans, 1761–62b, vol. 52, pp.196–201; Nevil Maskelyne's
       Autobiographical Notes, transcribed in Howse 1989, Appendix B,
       p.216; Howse 1989, p.33
80   'intensely black spot': Maskelyne, *Phil Trans*, 1761–62b, vol. 52,
       p.197
81   'degree of exactness': *Ibid.*, p.197
81   'principal observation of all': *Ibid.*, p.198
81   'more favourable opportunity': *Ibid.*, p.199
81   'I am afraid': *Ibid.*, p.199
81   'we have done all': *Ibid.*, p.201
81   Winthrop's day of transit in Newfoundland: Winthrop, *Phil Trans*,
       1764, vol. 54, pp.279–283; Winthrop 1761, p.9ff
81   North America's greatest mathematician: Brasch 1916, p.157
81   'most important phenomenon' and following quote: Winthrop Lectures,
       1769, quoted in Brasch 1916, p.166
81   'topic of conversation': Winthrop 1761, p.7
82   Winthrop's journey: *Ibid.*, pp.8–9
82   'towards the Sun-rising': *Ibid.*, p.9
82   'obliged to seek farther': *Ibid.*
82   'some distance': *Ibid.*, p.10
82   'swarms of insects' and following quote: *Ibid.*
82   'serene and calm': *Ibid.*, p.11
82   'most agreable Sight': *Ibid.*
82   'curious a spectacle': *Ibid.*
82   'Venus's Hill': *Ibid.*, p.12
82   waited for this 'solemn day': Kordenbusch 1769, p.46
83   Kordenbusch heard thunder: *Ibid.*, pp.60–61
83   'fear and hope': Johann Georg von Lori to Prosper Goldhofer, 6 June
       1761, Spindler 1959, p.411; Westenrieder 1784, p.75
83   Tobias Mayer's observations in Göttingen: *Göttingische Anzeigen von
       Gelehrten Sachen*, 1761–62, vol. 1, pp.57–58
83   'devoured by the clouds': Prosper Goldhofer to Johann Georg von
       Lori, 11 June 1761, Spindler 1959, p.428
83   'the unlucky clouds': Mayer, *Phil Trans*, 1764, vol. 54, p.163
83   'Jesus, there she is!': Johann Georg von Lori to Prosper Goldhofer,
       6 June 1761, Spindler 1959, p.411

83   'splendid' sight: *Ibid.*

83   'almost no hope': Eichholz in Halberstadt, *Das Neuste aus der anmuthigen Gelehrsamkeit*, vol. 11, 1761, p.421

83   fable about Venus: 'Warum die Vereinigung mit der Sonne nicht sichtbar gewesen. Eine Fabel', *Ibid.*, p.417ff

83   'a glimpse of Venus': Johannes Lulofs, 4 March 1762, JBRS, vol. 52, f.67

83   astronomers in Amsterdam: *Das Neuste aus der anmuthigen Gelehrsamkeit*, vol. 11, 1761, p.502

83   astronomers in Sweden: Wargentin, KVA Abhandlungen,1761, p.161ff

83   Rumovsky in Selenginsk: Kordenbusch 1769, pp.57–58; 31 May/11 June 1762, Protocols, vol. 2, pp.483–484

85   'almost despaired': Bliss, *Phil Trans*, 1761–62, vol. 52, p.174

85   transit day in London: Short, *Phil Trans*, 1761–62a, vol. 52, p.182; Bliss, *Phil Trans*, 1761–62, vol. 52, pp.175–176; Canton, *Phil Trans*, 1761–62, vol. 52, p.183; Dunn, *Phil Trans*, 1761–62, vol. 52, pp.189–193

85   Delisle's day of transit: 7 June 1761, *Gaceta de Madrid*

85   Château de Saint-Hubert: Le Monnier, *Histoire & Mémoires*, 1761, p.72

85   Delisle's mappemonde and proposal printed in *Novelle Letterarie*: this was Father Niccolo Maria Carcani, *Novelle Letterarie*, 1761, pp.280–285; Pigatto, 2004, p.75

85   Italian observers: Pigatto 2004, p.74ff; Spindler 1959, p.431; Zanotti, *Phil Trans*, 1761–62, vol. 52

85   'Madame Venus': Prosper Goldhofer to Johann Georg von Lori, 29 March 1761, Spindler 1959, p.389

86   projections of Venus: Eichholz in Halberstadt and Polack in Frankfurt, *Das Neuste aus der anmuthigen Gelehrsamkeit*, vol. 11, 1761, pp.422, 495

86   performance in Martin's shop: Millburn 1976, p.123

86   'entertained a considerable': *Boston Evening Post*, 7 September 1761; another officer timed the transit in Pondicherry but it was believed to be not exact enough, 22 April 1762, JBRS, vol.25, p.110

86   Bermuda wedding: Watlington 1980, p.185

86   'to amuse' the locals: Hellant, KVA Abhandlungen, 1761, p.180

86   'The more Observers': *Boston Evening Post*, 27 July 1761

7 How Far to the Sun

87   Le Gentil after transit: Le Gentil to Lanux, 16 July 1761, Le Gentil 1779 and 1781, vol. 2, p.750

87   Le Gentil missed Pingré: *Ibid.*

87   Pingré finished observations: Pingré, *Phil Trans*, 1761–62, vol. 52, pp.374–375

87   British attack in Rodrigues: Pingré to British Admiralty, 15 September 1761, reprinted in Woolf 1959, p.204ff; Pingré to Académie des Sciences, 19 September 1761, read on 30 January 1762, PV Académie, 1762, f.10ff; 29 June 1761, Pingré 2004, p.189ff

87   'half of our weapons': 29 June 1761, Pingré 2004, p.190

87   'not to molest': Pingré to British Admiralty, 15 September 1761, reprinted in Woolf 1959, p.205; 29 June 1761, Pingré 2004, p.191

88   English left Rodrigues: 3 July 1761, Pingré 2004, p.195
88   Houses were 'ravaged': Pingré to British Admiralty, 15 September 1761, reprinted in Woolf 1959 p.206
88   British pillaged Rodrigues: Pingré to British Admiralty, 15 September 1761, reprinted in Woolf 1959 p.206; July 1761, Pingré 2004, p.193
88   600 pounds of rice and flour: Pingré to British Admiralty 15 September 1761, reprinted in full in Woolf 1959, p.208; in his journal Pingré wrote that they had 1,200 pounds of rice and 400 pounds of flour because they found some more after the British had left, 1761, Pingré 2004, p.190
88   'the disgusting beverage': Pingré to British Admiralty, 15 September 1761, reprinted in Woolf 1959, p.207; Pingré to Académie des Sciences, 19 September 1761, read on 30 January 1762, PV Académie, 1762, f.10
88   'politeness and humanity': Pingré to British Admiralty, 15 September 1761, reprinted in Woolf 1959 p.207; see also July 1761, Pingré 2004, pp.196–197
88   captain took Pingré's observations: Pingré to Royal Society, 24 July 1761, read on 29 April 1762, JBRS, vol. 25, f.113ff
88   petty fights: August 1761, Pingré 2004, pp.200–201
88   Le Gentil decided to stay: Le Gentil (Ebeling) 1781, p.37
89   'require several years': Ibid.
89   'compensate' for disappointment: Ibid.
89   'wait for the Transit': Ibid.
89   observations 'as useful as possible': Le Gentil to Duc de Chaulnes, 6 September 1761, read on 30 January 1762, PV Académie 1762, f.19
89   'least bad observation': Ibid., f.20
89   'he lacks more than half': Ibid.
89   'take the waters': Planman to Wargentin, 11 June 1761, MS Wargentin, KVA, Centre for History of Science
89   'I got tired of': Planman to Mallet, 11 June 1761, quoted in Lindroth 1967, p.403
89   experiments on gravity: Howse 1989, p.37
90   Mason and Dixon continued observations: Mason and Dixon, Phil Trans, vol. 52, 1761–62, pp.378–394
90   For Mason and Dixon's travels between Cape and St Helena: Extract of the General Letter from St Helena dated 25 Jan 1762, RS MM/10, f.140; Howse 1989, p.37–38; Mason and Dixon, Phil Trans, 1761–62, vol. 52, p.393
90   Reports read: at RS 11 June 1761, JBRS, vol. 25, f.121ff; at Académie: 10, 13, 17, 20 and 27 June 1761, PV Académie 1761, ff.108–121; and in St Petersburg: 11 June 1761, Protocols, vol. 2, p.468
90   exchange of results: French ambassador reported Danish results: Lalande, Histoire & Mémoires, 1761b, pp.113–14; British ambassador in Constantinople: Porter, Phil Trans, 1761–62, vol. 52, p.226; 12 November 1761, JBRS, vol. 25, f.159; Italian observations to RS: Zanotti, Phil Trans, 1761–62, vol. 52, pp.399–414; 12 November 1761, JBRS, vol. 25, f.149 and 14 January 1762, JBRS, vol. 26, f.14; German observations to RS: Mayer, Phil Trans, 1764, vol. 54, pp.163–164; French observations to RS: 19 November 1761, JBRS, vol. 25,

f.159ff; 29 April 1762, JBRS, vol. 26, f.113ff; Munich, Bavarian monastry and Italian results: Franz Töpsl to Johann Georg von Lori, 23 July 1761, Spindler, 1959, p.432; reports arrived in St Petersburg: from Frisius and Zanotti, 20/31 August 1761, Protocols, vol. 2, p.470; from Asclepi and Ximenes, 1/12 February 1762, Protocols, vol. 2, p.479; Swedish results to Paris, St Petersburg and London: 1 July 1761, PV Académie, f.123, 31 August/11 September 1761, Protocols, vol. 2, p.471, 12 November 1761, JBRS, vol. 25, f.151, Wargentin, *Phil Trans*, 1761–62, vol. 52, p.215

91   German results published: this was Georg Christoph Silberschlag; published in the *Magdeburgischen Zeitung* on 13 June 1761; see also Donnert 2002, p.66

91   Italian results published: this was Giovanni Battista Audiffredi; Pigatto 2004, p.77

91   Turin results published: *Ibid.*, p.75

91   Padua results published: *Ibid.*

91   Lomonosov's results published: Meadows 1966, p.117

91   Boston papers: *Boston Evening Post*, 7 September 1761

91   captain took Pingré's results to RS: Pingré to RS, 24 July 1761, read on 29 April 1762, JBRS, vol. 2 5, f.113ff. Pingré sent a second report from Lisbon on 6 March 1762, Pingré, *Phil Trans*, 1761–62, vol. 52, p.371ff.

91   Pingré's rescue from Rodrigues: 6 September 1761, Pingré 2004, p.201ff.

91   Mason and Dixon's return: 22 April 1762, JBRS, vol. 25, f.100; Invoice Mason and Dixon, 7 April 1762, RS MM/10, f.143

92   Mason–Dixon Line and RS: Cope and Robinson 1951, p.55ff

92   Chappe collapsed: Fouchy 1769, p.167

92   'overwhelming weakness' and Chappe's return journey: *Ibid.*

92   'extremely tired, which': 21 November 1761, PV Académie 1761, f.206

92   Chappe's observations reach St Petersburg: 24 July/4 August 1761; 20/31 August 1761; 24 August/4 September 1761. On 22 January 1762 Chappe attended an Academy meeting in St Petersburg where he read his observations, 11/22 January 1761, Protocols, vol. 2, pp.469–471

92   Académie received Chappe's pamphlet: 5 May 1762, PV Académie 1762, f.189

92   Comments like 'doubtful': 12 November 1761, JBRS, vol. 25, f.154; Wargentin, KVA Abhandlungen, 1761, p.149

92   resembled a 'pear': Hirst, *Phil Trans*, 1761–62, vol. 52, p.398

93   'shape of a drop': Wargentin, KVA Abhandlungen, 1761, p.147

93   'tip of a rapier': *Ibid.*, p.150

93   'alternately dilated': Maskelyne, *Phil Trans*, 1761–62b, vol. 52, p.197

93   'stick to the Sun': Mr Dunthorn, 21 January 1762, JBRS, vol. 25, f.21

93   term 'black drop' first used: Lexell 1772, p.100

93   'blink of an eye': Röhl 1768, p.119

93   twenty-two seconds: Wargentin, KVA Abhandlungen, 1761, p.146

93   edge of sun 'trembling': Mr Dunthorn, 21 January 1762, JBRS, vol. 25, f.21

93   'exceedingly ill defined': Maskelyne, *Phil Trans*, 1761–62b, vol. 52, p.196

93   'vehement undulations': and following quote, Wargentin to RS, 12
      November 1761, JBRS, vol. 25, f.151

93   luminous ring: Hirst, *Phil Trans*, 1761–62, vol. 52, pp.397–398;
      astronomers in Stockholm, Wargentin, KVA Abhandlungen, 1761,
      p.181; astronomers in Uppsala, Wargentin, KVA Abhandlungen, 1761,
      pp.145– 147; Ferner in France, Ferner, *Phil Trans*, vol. 52, 1761–62,
      p.223; Dunn in Chelsea, Dunn, *Phil Trans*, vol. 52, 1761–62, p.192;
      Röhl in Greifswald, Röhl 1768, p.118; Silberschlag in Kloster Berge,
      Das Neuste aus der Anmuthigen Gelehrsamkeit, vol. 11, 1761, p.425

93   Silberschlag's observations: Kordenbusch 1769, pp.55–56

93   Lapland observations: Hellant, KVA Abhandlungen, 1761, p.181

93   Paris observations: Ferner, *Phil Trans*, 1761–62, vol. 52, p.223

93   London observations: Dunn, *Phil Trans*, 1761–62, vol. 52, p.192

93   astronomers' conclusions upon atmosphere on Venus: Wargentin, KVA
      Abhandlungen, 1761, p.146; Georg Christoph Silberschlag in *Das
      Neuste aus der Anmuthigen Gelehrsamkeit*, vol. 11, 1761, p.425;
      Dunn, *Phil Trans*, 1761–62, vol. 52, p.192, Benjamin Wilson to Torbern
      Olof Bergman, 14 December 1761, Göte and Nordström 1965, vol.
      1, p.419

94   Planman's times wrong: Planman to Wargentin, 14 June 1761, MS
      Wargentin, KVA, Center for History of Science

94   maybe telescope failed: Planman to Wargentin, 3 July 1761, MS
      Wargentin, KVA, Center for History of Science

94   'a whole minute': Planman to Wargentin, 25 July 1761, MS Wargentin,
      KVA, Center for History of Science

94   'suspected' his assistant: *Ibid.*

94   assistant noted wrong figures: *Ibid.*

94   Planman changed times: *Ibid.*

95   Wargentin worried about fluctuating Swedish data: Lindroth 1967,
      p.403

95   others changing numbers: Mallet to Planman, 9 July 1761, Nordenmark
      1939, p.179

95   'erred in recording': 13/24 September 1764, Protocols, vol. 2, p.525

95   'the slowness of the': Pingre to RS, 14 February 1764, Pingré, *Phil
      Trans*, 1764, vol. 54, p.159; Pingré to RS, 24 July 1761, read on 29
      April 1762, JBRS, vol. 25, f.115; Pingré to RS, 3 March 1762, Pingré,
      Phil Trans, 1761–62, vol. 52, p.371ff

95   'absurd consequences': James Short, 2 June 1763, JBRS, vol. 25, f.94

95   Delisle increasingly feeble: Anon., 'Éloge de M. de l'Isle', *Histoire &
      Mémoires*, 1768, p.182; entry Delisle, Gillispie 1970–80

95   his true 'heirs': Anon., 'Éloge de M. de l'Isle', *Histoire & Mémoires*,
      1768, p.182

95   Lalande biographical information: Alder 2002, p.82ff; entry Lalande,
      Gillispie 1970–80

95   'oilskin for insults': entry Lalande, Gillispie 1970–80

96   Lalande and comet in Paris: entry Lalande, Gillispie 1970–80; Alder
      2002, p.82

96   Lalande parallax: Lalande to RS, read on 25 February 1762 and
      29 April 1762, JBRS, vol. 25, ff.58–59, 116ff

96   Pingré parallax: Pingré, *Phil Trans*, 1764, vol. 54, p.152

96   Short parallax: Short, *Phil Trans*, 1761–62b, vol. 52, p.618

96   Planman and parallax obsession: Nordenmark 1939, p.181ff
96   Planman and parallax value: Planman, KVA Abhandlungen, 1763, pp.135, 139
96   'strong calculator': Wargentin to Planman, 18 March 1763, quoted in Nordenmark 1939, p. 182
96   'I am quite at a loss': *Ibid.*
96   parallax ranged between 8" and 10": Lalande to Leonard Euler, reported by Leonard Euler to Müller, 26 June 1762, Juskevic and Winter 1959–76, vol. 1, p.194
96   'There was no reason': *Ibid.*
96   'but I fear there is not': Benjamin Wilson to Torbern Olof Bergman, 14 December 1761, Göte and Nordström, 1965, vol. 1, p.419
97   comparing these observations: Wargentin to RS, 7 August 1761, Wargentin, *Phil Trans*, 1761–62, vol. 52, p.215
97   parallax ranged from 8"28 to 10"6: Woolf 1959, p.192
97   parallax distances: Maor 2004, p.92
97   battle between Rumovsky and Popov: 5/16 March 1764, Protocols, vol. 2, p.510ff (and entries over the next two months)
97   'obtain the judgment': 8/19 March 1764, *Ibid.*, p.513
97   Popov promised to recalculate: 29 March/9 April 1764, *Ibid.*, p.514
97   'seek his remedy': 2/13 April 1764, 1764, *Ibid.*, p.514
97   'useless and harmful': 30 April/11 May 1764, *Ibid.*, p.515
97   'seem almost without': 23 August/3 September 1764, *Ibid.*, p.524
97   Popov gave up: 13/24 September 1764, *Ibid.*, p.525
97   'truth of these observations': 26 January/6 February 1764, *Ibid.*, p.510
98   'a mistake of one minute': Short, *Phil Trans*, 1763, vol. 53, p.301
98   'uncertain and even': Pingré, *Histoire & Mémoires*, 1765, p.23; see also Woolf 1959, p.147
98   'prefer his own observations': Hornsby, *Phil Trans*, 1763, vol. 53, p.492
98   Audiffredi humiliated: Pigatto 2004, p.81
98   'astronomers' disputes': Audiffredi, quoted in *Ibid.*, p.81
98   Audiffredi and parallax: *Ibid.*, p.83
98   'These considerations, I say': Benjamin Wilson to Torbern Olof Bergman, 14 December 1761, Göte and Nordström 1965, vol. 1, p.419
98   'a definite decision': Pingré, *Phil Trans*, 1764, vol. 54, p.152, see also Lalande to RS, read on 29 April 1762, JBRS, vol. 25, f.118; Benjamin Franklin to John Winthrop, 23 December 1762, BF online
98   'we must wait for': Maskelyne to RS, 29 June 1761, read on 5 November 1761, JBRS, vol. 25, f.138

8 A Second Chance

101   George III and science: Black 2006, pp.181–184
101   'love of sciences': Samuel Johnson on George III, quoted in Black 2006, p.183
101   Carlos III and science: Engstrand 1981, p.6
101   Louis XV observed first transit: Le Monnier, *Histoire & Mémoires*, 1761, pp.72–76
101   Queen Louisa Ulrika at first transit: Lindroth 1967, p.402
101   Christian VII and RS: 1 September 1768, JBRS, vol. 27, f.117
101   Christian VII and Académie: Hahn 1971, p.74

101   Catherine and smallpox: Voltaire to Catherine, 26 February 1769; Lentin 1974, p.19

102   George III's observatory: Scott 1885, p.42

103   'The knowledge of': Hornsby, *Phil Trans*, 1765, vol. 55, p.327
Lalande's mappemonde presented in March 1764: Lalande, *Histoire & Mémoires*, 1764, pp.122–124; Woolf 1959, p.151

103   Lalande in London: JBRS, vol. 25, ff.43, 50, 64, 92

103   because of 'vapours': Hornsby, *Phil Trans*, 1765, vol. 55, p.332

103   transit Uppsala 1761: *Ibid.*

104   'any other place near': Ferguson, *Phil Trans*, 1763, vol. 53, p.30; see also for predictions Hornsby, *Phil Trans*, 1765, vol. 55

104   best viewing in South Pacific and following: Hornsby, *Phil Trans*, 1765, vol. 55, p.332ff

105   size of Pacific Ocean: McLynn 2011, p.72

105   'any such islands': Hornsby, *Phil Trans*, 1765, vol. 55, p.336

105   difference to Torneå at least twenty minutes: *Ibid.*, pp.336–338

105   Mexico as alternative: *Ibid.*, pp.339–340

105   'we have advantages': Pingré, 25 February 1767, PV Académie 1767, f. 49; see also Pingré, *Histoire & Mémoires*, 1767, pp.105–109

105   Pingré turned to old travel accounts: Pingré, *Histoire & Mémoires*, 1767, p.107

106   suggested Marquesas Islands: Pingré, 25 February 1767, PV Académie, f.50; Pingré, *Histoire & Mémoires*, 1767, p.108

106   'of sweet character': Pingré, 25 February 1767, PV Académie, f.50

106   'posterity must reflect': Hornsby, *Phil Trans*, 1765, vol. 55. p.344

106   'A commercial nation': *Ibid.*

106   'much more safety': Le Gentil (Ebeling) 1781, p.42

107   'it was time to think': *Ibid.*, p.44

107   Le Gentil's plans and calculations: *Ibid.*, p.45

107   Le Gentil wrote to Académie: Le Gentil to Duc de Chaulnes, 10 January 1766, read on 11 June 1766, PV Académie 1766, f.199

107   Spanish vessel arrived in Mauritius: Le Gentil (Ebeling) 1781, pp.45–46

107   Le Gentil's departure from Mauritius: *Ibid.*, p.46

107   'one or more astronomical': 5 June 1766, CMRS, vol. v, f.145

108   Chappe's return from Russia: Fouchy 1769, p.168

108   Chappe's life between two transits: Armitage 1954, p.287

108   tested Berthoud's timepieces: *Ibid.*, p.286

108   Chappe's book: Chappe's *Voyage en Sibérie, fait par ordre du roi en 1761* (1768); on 31 August 1768, the Académie approved the publication, PV Académie 1768, ff.202–204

108   transit in American West: Hornsby, *Phil Trans*, 1765, vol. 55, pp.339–340

109   RS and Spanish ambassador: Prince de Masserano to Charles Morton, 22 August 1766, read on 25 November 1766, CMRS, vol. v, f.152; see also Nunis 1982, p.44ff

109   'some spanyards': Prince de Masserano to Charles Morton, 22 August 1766, read on 25 November 1766, CMRS, vol. v, f.152; see also Nunis 1982, p.44

109   Carlos III expelled Jesuits: Nunis 1982, p.44

109   'unique observation': Charles Morton to Prince de Masserano, 15 May 1767, *Ibid.*, p.45

109 'No permission will be': Council of Indies to Prince de Masserano, 13 July 1767, *Ibid.*, p.46

109 'good astronomers and': *Ibid.*

109 'His Majesty . . . will': *Ibid.*

110 'horrendous storm': Le Gentil (Ebeling) 1781, p.50; see also Le Gentil to Lanux, 1 September 1766, Le Gentil 1779 and 1781, vol. 2, p.788

110 prayers and 'alms': Le Gentil to Lanux, 1 September 1766, Le Gentil 1779 and 1781, vol. 2, p.788

110 vow to calculate longitude: *Ibid.*

110 'determined to suffer': *Ibid.*, p.760

110 'too far east': Le Gentil (Ebeling) 1781, p.57

110 weather in Manila and Pondicherry: *Ibid.*

110 'Flee this cruel': Le Gentil to Lanux , 1 October 1768, Le Gentil 1779 and 1781, vol. 2, p.792. From *The Aeneid* (3.44): 'Heu! fuges crudeles terras, fuge litus avarum'.

9 Russia Enters the Race

111 'the utmost care': Catherine the Great to Vladimir Orlov, 3/14 March 1767, Crosby 1797, p.73

111 'some resemblance to': William Richardson, quoted in Wolff 1994, p.84

111 danger of 'relapse': *Ibid.*, p.84

111 'Russia in Asia': *Ibid.*, p.23

111 'oriental empire': *Ibid.*, p.84

112 'Half of Russia': William Richardson in 1768, Putnam 1952, p.147

112 Russia after Seven Years' War: Madariaga 1990, p.24

112 Catherine contacted Voltaire: Lentin 1974, p.10

112 'governed the whole': Goethe, quoted in Gorbatov 2006, p.72

112 'It was certainly Voltaire': Catherine, quoted in Lentin 1974, p.13

112 Voltaire delighted to be inspiration: Lentin 1974, p.16

112 'family matters in which': Voltaire to Mme du Deffand, 18 May 1767, *Ibid.*, p.14

112 Catherine and also Catherine and Diderot's library: *Ibid.*, p.222; Madariaga 1990, p.96

112 'all the men of letters': Voltaire to Catherine the Great, November 1765, Lentin 1974, p.38

112 'Russia is a European power': *Ibid.*, p.29

112 Catherine and Legislative Commission: Dixon 2010, pp.156–157; Madariaga 1990, p.28

113 publication of transit observation in Selenginsk: 26 August/6 September 1762, Protocols, vol. 2, p.487

113 Aepinus at Court: Stählin to Leonard Euler, 2/13 April 1765, Juskevic and Winter 1976, vol. 3, p.232

113 tour along Volga: Dixon 2010, p.158

113 'most advantageously situated': Catherine the Great to Vladimir Orlov, 3/14 March 1767, Crosby 1797, p.73

113 Catherine's letter read at Academy: 16/27 March, Protocols, vol. 2, p.595

113 Academy's early interest in second transit: 21 October/1 November and 1/12 December 1764, Protocols, vol. 2, pp.528, 530

113 'training of some younger': 2/13 October 1766, Protocols, vol. 2, p.574

114   discussions at the Academy: 16/27 March, 19/30 March, 23 March/3 April 1767, Protocols, vol. 2, pp.595–596

114   Academy's letter to Catherine and following description: Academy to Catherine and Count Orlov, 26 March/6 April 1767, Protocols, vol. 2, pp.596–599

114   only twenty locations determined: Bacmeister 1772, vol. 1, p.51

114   four expeditions: 26 March/6 April 1767, Protocols, vol. 2, p.596

114   'all the possible effort': Academy to Catherine and Count Orlov, 26 March/6 April 1767, Protocols, vol. 2, p.598

114   'basics of mathematics': *Ibid.*, p.597

114   suggested instruments: *Ibid.*, p.598

114   observatories for expeditions: *Ibid.*

114   Academy wrote letters re expeditions: 20 April/1 May, 25 June/6 July 1767, 13/24 August 1767, Protocols, vol. 2, pp.600, 607, 610; Stepan Rumovsky to James Short, 23 October 1767, RS L&P, Decade V, f.1

114   letters from Germany: 20 April/1 May 1767 and 11/22 May 1767 and 10/21 August 1761, Protocols, vol. 2, pp.600, 603, 609

114   letters from France: Lalande to Russian Academy, 1 June 1767, Moutchnik 2006, p.183

115   'further reports from': Bacmeister 1772, vol. 1, p.46

115   'foggy' in the north: *Ibid.*

115   Catherine doubled the viewing stations: *Ibid.*

115   'immediately granted': *Ibid.*, p.47

115   Catherine ordered Aepinus: 2/13 July 1767, Protocols, vol. 2, pp.607–608

115   Catherine dispatched Orlov to St Petersburg: Bacmeister 1772, vol. 1, p.48

115   Aepinus and timber: 2/13 June 1768, Protocols, vol. 2, p.641

115   'dispatched' to Kola: Bacmeister 1772, vol. 1, p.48

115   little cabins should be constructed in advance: Bacmeister 1772, vol. 1, p.53

115   'low morals': 22 October/2 November 1767, Protocols, vol. 2, p.623

115   foreign scientists to Russia: 22 October/2 November 1767, Protocols, vol. 2, p.623; from Leipzig Wolfgang Ludwig Krafft: 11/22 May 1767, Protocols, vol. 2, p.603, from Göttingen Georg Moritz Lowitz: Lowitz to Leonard Euler, 26 July 1767, Juskevic and Winter 1959–1976, vol. 3, p.211; 10/21 August and 13/24 August and 17/28 August 1767, Protocols, vol. 2, pp.609–610; Lowitz to Euler, 30 September 1767, Pfrepper and Pfrepper 2009, p.108; from France: Lalande to Russian Academy, 1 June 1767, Moutchnik 2006, p.183; from Geneva Jacques André Mallet and his assistant Jean-Louis Pictet: 5/16 October 1767, Protocols, vol. 2, p.619.

116   letter from James Short: 8/19 October 1767, Protocols, vol. 2, p.621

116   'already began to doubt': Stepan Rumovsky to James Short, 23 October 1767, read on 14 January 1768, RS L&P, Decade v, f.1

116   'delivered us from a': *Ibid.*

116   'unpredictability' of weather: *Ibid.*

116   report to Catherine: 19/30 October and 22 October/2 November 1767, Protocols, vol. 2, pp.621–624

116   scientists to collect natural history objects: Bacmeister 1772, vol. 1, p.48; 5/16 October 1767, Protocols, vol. 2, p.620

116 astronomers to be accompanied by scientific teams: 22 October/2 November 1767, Protocols, vol. 2, p.623; Bacmeister 1772, vol. 1, pp.89–90

117 expedition teams: 22 October/2 November 1767, Protocols, vol. 2, p.623

117 Lowitz and canal: Pfrepper and Pfrepper 2009, p.109

117 'whole of Europe is': Voltaire to Catherine, 30 September 1767, Lentin 1974, p.20

117 'at the expense of': 26 November 1767, *New York Journal*; 4 December 1767, *Connecticut Journal*; 2 January 1768, *Providence Gazette*

117 Russian newspaper: Bacmeister 1772, vol.1, p.47

117 Russian team to Yakutsk: Islenieff left on 11/22 February 1768; Bacmeister 1772, vol. 1, p.49; Lowitz's delay: 29 February/11 March 1768, Protocols, vol. 2, p.632; see also Pfrepper and Pfrepper 2009, p.108

117 Lowitz arrived in St Petersburg: Pfrepper and Pfrepper 2004, p.171

118 additional 10,000 rubles: Bacmeister 1772, vol. 1, p.50

118 Short's illness in June 1768: Turner 1969, p.94

118 achromatic telescope: Moutchnik 2006, p.155

118 'a very happy invention': Mayer, *Phil Trans*, 1764, vol. 54, p.161

118 Russian order of instruments: Stepan Rumovsky to James Short, 23 October 1767 (Short's list of instruments was on the back of the letter), RS L&P, Decade V, f.1

119 Short determined to finish order: Turner 1969, pp.94, 105

119 delays to Yakutsk expedition: 30 June/11 July 1768, Protocols, vol. 2, p.644. This was a letter from Islenieff posted on 26 March/6 April 1768.

119 decision to send scientists in advance: Wendland 1992, vol. 1, p.97; Bacmeister 1772, vol. 1, p.94

119 Short's instruments arrived: 6/17 October 1768, Protocols, vol. 2, p.653

119 instruments from France: 14/25 July, 24 October/4 November and 5/16 December 1768, Protocols, vol. 2, pp.646, 655, 659

119 'practise their usage': 6/17 October 1768, Protocols, vol. 2, p.653

119 destinations and teams: Georg Moritz Lowitz and his Russian assistant to Guryev; Wolfgang Ludwig Krafft and his team to Orenburg; Christoph Euler, who was the son of one of the many German scientists at the Academy in St Petersburg, to Orsk. The Russian Stepan Rumovsky and his team to Kola, Jacques André Mallet to Ponoy and Jean-Louis Pictet to Umba.

119 instructions read: 16/27 January 1769, Protocols, vol. 2, p.663; the astronomers received the written instructions on the day after the visit to the Winter Palace.

119 not 'constrained': Bacmeister 1772, vol. 1, p.56

119 'very latest' and other instructions: Bacmeister 1772, vol. 1, p.56

120 audience with Catherine: Inochodcev to Kästner, 3 February 1769, transcribed in Pfrepper and Pfrepper 2009, p.110; Bacmeister 1772, vol. 1, p.52

120 description of that winter in St Petersburg: William Richardson in winter 1768–69, Putnam 1952, p.143

120 washerwomen on Neva: Wolff 1994, p.37

120 'atonishing swiftness': William Richardson in winter 1768–69, Putnam 1952, p.145

120 people sliding on trays: Elizabeth Dimsdale in 1781, Cross 1989, p.41

120 'in Asiatic costume': Diderot quoted in Wolff 1994, p.22

120 hardened by 'clotted ice': William Coxe quoted in Wolff 1994, p.37

120 'age of barbarism': Diderot quoted in Wolff 1994, p.22

120 St Petersburg a combination of Asia and London/Paris: Wolff 1994, p.37

121 fires for servants: Coxe 1784, vol. 2, p.76

121 'gaudy butterflies': William Richardson in winter 1768–69, Putnam 1952, p.144

121 Catherine's art collection: Rounding 2007, p.217

121 'foreign gentlemen': William Coxe, Rounding 2007, p.345

121 ceremonial kiss: 18/29 January 1769, Catherine's Court Journals 1769, Sobolevskii 1853–67

121 Catherine still beautiful: Madariaga 1990, p.205

121 'expressive of scrutiny': William Richardson in August 1768, Putnam 1952, p.145; see also Madariaga 1990, p.4

121 'brilliant' conversation: George Macartney, 1766, quoted in Madariaga 1990, p.205

121 kissing Catherine's hand: Inochodcev to Kästner, 3 February 1769, transcribed in Pfrepper and Pfrepper 2009, p.110

10 The Most Dangerous Voyage of All

122 'Transit Committee': 12 November 1767, CMRS, vol. v, f.172

122 'certainty and excellency': Maskelyne to Royal Society, 14 May 1762, read on 20 May 1762, JBRS, vol. 25, f.134

122 'great benefit': Maskelyne to East India Company, read on 24 June 1762, JBRS, vol. 25, f.172

123 Maskelyne Astronomer Royal: Howse 1989, pp.54–59

123 'you must', 'you will': Maskelyne to Charles Mason, 29 November 1760, RGO 4/184, Letter 1

123 decision to send British expeditions: 17 November 1767, CMRS, vol. v, ff.176–177

124 'except we learn that': 17 November 1767, CMRS, vol. v, ff.176 and Committee, 19 November 1767, transcribed Beaglehole 1999, vol. 1, p.511

124 'government be petitiond': *Ibid.*, f.177

124 'about Xmas 1768': *Ibid*

124 Maskelyne's observing suit: Howse 1989, pp.100–101

124 'small invisible man': Johan Henrik Lidén to Fredrik Mallet, 10 July 1769, Heyman 1938, p.281

124 Chappe presented South Sea expedition proposal: 14 November 1767, PV Académie 1767, no folio number (but inserted after f.242)

125 'several centuries': *Ibid.*

125 RS progress report: 3 December 1767 CMRS, vol. v, ff.181–200

125 'to know at what places': *Ibid.*, f.199

125 candidates for expeditions: 18 and 22 December 1767, CMRS, vol. v, f.227ff

125 asked assistance from East India Company: 3 December 1767 and 21 January 1768, CMRS, vol. v, ff.200, 265

125 letter from the Academy in St Petersburg: 14 January 1768, JBRS, vol. 27, f.6

125 map of Mason–Dixon Line: Mason 1969, p.21

125 newspaper advertisement: for example by Benjamin Martin in *Gazetteer and New Daily Advertiser*, 15 January 1768

126 Maskelyne's instructions: printed and presented to the RS, 5 May 1768, JBRS, vol. 27, f.87

126 'flying clouds': Maskelyne 1768a, p.6

126 'attentive to what is to pass in the Heavens': Benjamin Franklin to Jean-Baptiste LeRoy, 14 March 1768, BF online

126 drafting petition to king: 14 January 1768 and 15 February 1768, CMRS, vol.v, f.262, 282ff

126 Le Gentil leaving Manila: Le Gentil (Ebeling) 1781, pp.60–61

126 'Nation upon Earth': Memorial RS, 15 February 1768, transcribed Beaglehole 1999, vol. 1, p.604 and CMRS, vol. v, p.282ff

126 'justly celebrated in': *Ibid.*

126 observatories to Hudson Bay: 28 January 1768, CMRS, vol. v, f.275

127 portable observatories: 28 January, 11 February, 15 February, 10 March, 21 April 1768, CMRS, vol. v, ff.277–278, 281–282, 289, 295

127 candidate for Hudson Bay expedition: 3 December 1767, CMRS, vol. v, f.198

127 William Wales biographical information: Wales 2004, p.28

127 *Nautical Almanac*: Howse 1989, p.85ff

128 Hudson's Bay Company and expedition: 22 December 1767 and 28 January 1768, CMRS, vol. v, ff.231, 274ff

128 Maskelyne and portable observatory: 10 March 1768, CMRS, vol. v, f.289

128 space for observatory onboard: 28 January 1768, CMRS, vol. v, f.276

128 'His Majesty has been': Secretary of the Admiralty to Navy Board, 5 March 1768, transcribed Beaglehole 1999, vol. 1, p.605

128 possible ships: Letters between Admiralty and Navy Board, 8 March, 10 March, 21 March, 22, March, 27 March, 29 March 1768, transcribed Beaglehole 1999, vol. 1, pp.605–606

128 'stow the quantity': Navy Board to Admiralty, 21 March 1768, *Ibid.*, p.605

128 King George III granted £4,000: 24 March 1768, CMRS, vol. v, p.290

128 search for the *Endeavour*: Navy Board to Admiralty Secretary, 21 March and 29 March 1768; Admiralty Secretary to Navy Board, 21 March, 1768, transcribed Beaglehole 1999, vol. 1, pp.605–606.

128 *Endeavour* refitted in Deptford: 29 March and 18 May 1768, *Ibid.*, pp.606, 608

128 James Cook in charge of *Endeavour*: 12 April 1768, *Ibid.*, p.607

128 James Cook biographical information: McLynn 2011; Aughton 2003

129 Cook as observer: 5 May 1768, CMRS, vol. v, f.299

129   Charles Green as astronomer on *Endeavour*: *Ibid.*

129   Charles Green biographical information: Beaglehole 1999, vol.1, p.cxxxiv; Morris 1980 and 1981, vol. 3, p.92 and vol. 4, p.102

129   'go to the Southward': 18 December 1767, CMRS vol. v, f.227; see also 18 December 1767, Beaglehole 1999, vol. 1, p.512

129   salary of £100: 5 May 1768, CMRS, vol. v, f.299

129   Cook one-off payment of £100: McLynn 2011, p.70

129   Maskelyne wrote instructions: 19 May 1768, CMRS, vol. v, f.305; Maskelyne's instructions are reprinted in Kaye 1969, pp.11–13; Maskelyne observatory, 12 May 1768, CMRS, vol. v, f.301

129   'without any loss of time': Kaye 1969, p.12

129   Wales stored observatory and instruments onboard: 21 April 1768, 5 May 1768, CMRS, vol. v, ff.295, 301

129   Wales midnight visit to Greenwich: 29 and 30 May 1768, Wales 1770, p.100

130   Wales's wife giving birth: Wales 2004, p.29

130   'Swims too much': Cook to Navy Board, 30 June 1768, Beaglehole 1999, vol. 1, p.615

130   crew and spirit allowance: 26 and 27 May, 2–3 June 1768, *Ibid.*, pp.610–611

130   Samuel Wallis and Tahiti: Carter 1988, p.64; Beaglehole 1999, vol. 1, p.513

130   riots and strikes in London: McLynn 2011, p.85

130   'Surgeons necessarys': Cook to Navy Board, 6 July 1768, Beaglehole 1999, vol. 1, p.616

130   Cook ordered guns: Cook to Admiralty Secretary, 5 July 1768, *Ibid.*

130   'a Machine for sweetening': 10 June 1768, Navy Board Warrant, *Ibid.*, p.612

130   'portable soup': Cook to Admiralty Secretary, 23 June 1768, *Ibid.*, p.614

130   Cook ordered salt: Cook to Victualling Board, 11 July 1768, *Ibid.*, p.617

130   methods to fight scurvy: Cook to Sick and Hurt Board, 11 July 1768, *Ibid.*, p.618

130   Cook ordered surveying instruments: Cook to Admiralty Secretary, 8 July 1768, *Ibid.*, p.617

131   Cook complained about lame cook: Cook to Navy Board, 13 June 1768, *Ibid.*, p.613

131   'this man hath had': Cook to Navy Board, 16 June 1768, *Ibid.*, p.614

131   one-handed cook remained on *Endeavour*: Navy Board to Cook, 17 June 1768, *Ibid.*

131   'little Service': Cook to Navy Board, 16 June 1768, *Ibid.*

131   'utmost patience': Earl of Morton, 10 August 1768, *Ibid.*, p.514

131   primary object of the voyage: *Ibid.*, p.516

131   'a Continent in the': *Ibid.*, p.516

131   departure of *Endeavour*: 25 August 1768, Cook Journal

131   ninety-four men: Aughton 2003, p.20

132   Provisions on the *Endeavour*: Victualling Board Minutes, 13 June 1768, Beaglehole 1999, vol. 1, p.613; Aughton 2003, p.10

132 Green already on board: 7 August 1768, Anonymous Fragment, Beaglehole 1999, vol. 1, p.550

132 Joseph Banks's provisions: Joseph Ellis to Carl Linnaeus, 19 August 1768, Smith 1821, vol.1, p.231; see also Wulf 2008, pp.175-178; Carter 1988, p.71

132 'It almost frighten[s] me': Joseph Banks to William Philip Perrin, 16 August 1768, Chambers 2000, p.1

132 Banks's £10,000 bill: Ibid., p.2; Joseph Ellis to Carl Linnaeus, 19 August 1768, Smith 1821, vol. 1, p.231

132 'first man of Scientifick': Joseph Banks to Edward Hasted, February 1782, Dawson Turner Collection, Natural History Museum, London, vol. 2, f.97

132 changes to cabins: 17 August 1768, Cook Journal

132 Cook's wife giving birth: McLynn 2011, p.93

132 'excellent health and': 25 August 1768, Banks Journal

11 Scandinavia or the Land of the Midnight Sun

133 'I do believe, this': Wargentin to Planman, 4 July 1766, quoted Nordenmark 1939, p.184

133 'the only place': Ibid.

133 income Academy: Wargentin to Planman, 4 July 1766, Nordenmark 1939, p.184

133 'everlasting harm and': Ibid.

133 Wargentin biographical information: Frängsmyr 1989, pp.7-8, 50-56; Lindroth 1952, pp.105-110

134 'in the desire to: Wargentin quoted in McCellan 1985, p.171

134 Wargentin reports: 14 January 1767, KVA protocols, pp.801-802

134 discussions at Academy: Ibid., p.802

134 matter of national honour: Ibid.

134 'science, trade and seafaring': Petition to King Adolph Frederick, 14 January 1767, reprinted in Nordenmark 1939, p.375

134 'There was no other': Ibid., p.374

135 king granted funds: 29 January 1767, Nordenmark 1939, p.185

135 Wargentin suggested expedition: Ibid., p.188; Lindroth 1967, p.405

135 Mallet biographical information: Nordenmark 1946

135 Mallet deeply melancholic: Ibid., p.20

135 Mallet struggled: Ibid., p.10

135 taste for metropolitan life: Ibid., p.11ff

135 'plagued less by melancholy': Wargentin to Strömer, 22 May 1759, Ibid., p.17

135 Mallet watched first transit in Uppsala: Wargentin, KVA Abhandlungen, 1761, pp.143-151

135 have himself 'hanged': Mallet to Planman, 9 July 1761, Nordenmark 1939, p.179

135 'no horse would pass': Ibid.

135 Mallet gloomy and impatient: Nordenmark 1946, p.20

136 'I am incapable of': Mallet to Wargentin, no date, Ibid., p.23

136 Planman to Kajana: Nordenmark 1939, p.188; Wargentin to Mallet, 13 April 1767, Heyman 1938, p.274

136 Planman and parallax: Nordenmark 1939, p.181ff

136   Wargentin's report read in St Petersburg: 10/21 September 1767, Protocols, vol. 2, p.618

136   Rumovsky sent Swedish news to RS: Stepan Rumovsky to James Short, 23 October 1767, RS L&P, Decade V, f.1

136   Maskelyne was relieved: Nordenmark 1939, p.186

136   'One or other of these Stations': Stepan Rumovsky to James Short, 23 October 1767, RS L&P, Decade V, f.1

136   exert 'their utmost': James Short to Stepan Rumovsky, 21 January 1768, RS L&P, Decade V, f.1

136   Mallet received more than half of funds: 10 February 1768, KVA Protocols, p.835

137   'to determine the principal': Admiralty to Mallet, 12 April 1768, Nordenmark 1946, p.70; 20 April 1768, KVA Protocols, p.839

137   Mallet studied maps: Nordenmark 1946, p.70

137   three members to prepare instruments: 20 April 1768, KVA Protocols, p.839

137   'indescribably satisfied': Planman to Wargentin, 14 July 1768, MS Wargentin, KVA Center for History of Science

137   Mallet's departure from Uppsala: Nordemark 1939, p.188; Lindroth 1967, p.406

137   'deplorable' place: Mallet to Wargentin, 25 September 1768, Nordenmark 1946, p.72

137   'courage and perseverance': Ibid.

137   'I should have given up': Ibid.

137   downhearted letter to Wargentin: Ibid.

137   Christian VII's invitation to Hell: Aspaas, 2008, pp.10, 15

138   Maskelyne's pleas to Wargentin: Maskelyne to Wargentin, 5 January 1768, Nordenmark 1939, p.186

138   Christian VII ordered assistance in Vardø: 1 June 1768, Sajnovics Journal, Littrow 1835, p.100

138   'very much flattered': 1 September 1968, JBRS, vol. 27, f.117

138   Christian VII chose Jesuit priest: Hansen and Aspaas 2005, p.6

138   Maximilian Hell biographical information: Hamel *et al.* 2010, p.191ff

138   Hell and science and religion: *Ibid.*, p.196

139   Hell's departure: 28 April 1768, Sajnovics Journal, Littrow 1835, p.87; Hansen and Aspaas 2005, p.7

139   Hell's equipment and luggage: Aspaas 2008, p.10

139   meeting scientists: for example at the observatory in Prague and the astronomer Heinsius in Leipzig, 2 May 1768 and 13 May 1768, Sajnovics Journal, Littrow 1835, pp.87, 89

139   ascetic Hell: Hamel *et al.* 2010, p.196

139   Sajnovics's comments on Dresden: 8 May 1768, Sajnovics Journal, Littrow 1835, p.89

139   Sajnovics's comments on Hamburg: 24 May 1768, *Ibid.*, pp.93-94

139   roads between Hamburg and Lübeck: 25 May 1768, *Ibid.*, p.96

139   'wretched' cart: *Ibid.*

139   mercury floating in their clothes: 2 May 1768, *Ibid.*, p.87

140   'miserable' tavern: 30 May 1768, *Ibid.*, p.98

140   'I'm pleased . . . that': 1 June 1768, *Ibid.*, p.100

140   Catherine's astronomers to Arctic Circle: Swiss astronomers Jacques

André Mallet and his assistant Jean-Louis Pictet as well as Stepan
Rumovsky with an assistant.
140 expedition North Cape: 21 July 1768, CMRS, vol. v, f.328
140 King George at Royal Observatory: 2 June 1768, 'Observations of
Transits. A Working Copy of Transit Observations of the Major Stars,
7 May 1765–18 July 1771', RGO 4/3, p.121

12 The North American Continent
141 Wales's departure: Wales 1770, p.100
141 Spanish refusal to allow French into South Sea: Woolf 1959, p.157
141 Letters between Spain and France: letters between Don Georges Juan
and La Condamine in Woolf 1959, p.157, as well as discussions in
the Académie in Paris on 9 and 23 December 1767, PV Académie
1767, ff.288–289, 297–298 and on 9 January 1768, 23 April 1768,
PV Académie 1768, ff.1, 64
141 French preparations for California: 9 January, 23 April, 23 August
1768, PV Académie 1768, ff.1, 64, 189
142 Delisle 'attracted' to Ottoman princess: Delambre, *Histoire &
Memoires*, 1817, p.87
142 'could not part with her': *Ibid.*
142 Spanish expedition: Spanish Instructions, 27 April 1768, Nunis 1982,
pp.113–117
142 geographical positions: *Ibid.*, p.116
142 'never to separate themselves': *Ibid.*, p.114
142 meeting APS at State House: 21 June 1768, Minutes APS, *Proceedings
APS*, 1885, vol. 22, p.15
142 'Promoting Useful Knowledge': A Proposal for Promoting Useful
Knowledge, 14 May 1743, BF online
143 'very idle Gentlemen': 15 August 1745, Benjamin Franklin to
Cadwallader Colden, BF online
143 'degeneracy of America': Jefferson 1982; Cohen 1995, pp.72–79; Pauly
2007, pp.20–32
143 'man of genius': Jefferson 1982, p.64; Jefferson quoted Abbé Raynal
143 Reports on British expeditions: 10 June 1768, *Boston Post Boy*
143 Reports on Russian expeditions: 26 November 1767, *New York Journal*
143 'Much depends on this': 19 April 1768, Minutes APS, *Proceedings
APS*, 1885, vol. 22, p.14
143 'the first drudgery of': A Proposal for Promoting Useful Knowledge,
14 May 1743, BF online
143 Philadelphia: Greene 1993, p.87; Weigley 1982, p.68ff
144 Rittenhouse biographical information: Rufus 1928, pp.506–513; Rush
1796
144 carving astronomical constellations: Rufus 1928, p.506; Rush 1796,
p.8
144 predictions based on Rittenhouse's calculations: 21 June 1769, Minutes
APS, *Proceedings APS*, 1885, vol. 22, p.15
144 'the Beginning and a': 19 April 1768, *Ibid.*, p.13
144 'the necessary Preparations': 21 June 1769, *Ibid.*, p.15
144 APS learned about orrery: 22 March 1768, *Ibid.*, p.12
144 Rittenhouse's orrery: Cohen 1985, pp.80–85
145 'approached nearer its Maker': Jefferson 1982, p.64

145   BF 'Snuffling' air: Benjamin Franklin to Erasmus Darwin, 1 August 1772, quoted in Uglow 2002, p.238
145   BF and Gulf Stream: Chaplin 2006, pp.196–199
145   'America has sent us': David Hume to Benjamin Franklin, 10 May 1762, BF online
145   BF member of RS Council: 8 December 1760, CMRS, vol. iv, f.277
145   BF informed colonists about first transit: Benjamin Franklin to William Coleman, 12 October 1761, BF online
145   transit information to BF: Benjamin Franklin to John Winthrop, 23 December 1762, BF online
145   BF again elected to RS Council: 8 December 1766, CMRS, vol. v, f.157
145   BF and prospective candidates: 18 December 1767, *Ibid.*, f.227ff
146   BF and Hudson Bay: 22 December 1767, CMRS, *Ibid.*, ff.230–231
146   BF and petition to king: 14 January 1768, JBRS, vol. 27, f.6 and 14 January 1768, CMRS, vol. v, f.262
146   BF and portable observatories: 11 February 1768, CMRS, vol. v, f.281
146   BF and lists of instruments: 28 January 1768, *Ibid.*, ff.277–278
146   BF meeting Cook and Green: 5 May 1768, *Ibid.*, f.299ff
146   Winthrop ordered instruments from Short: Benjamin Franklin to John Winthrop, 2 July 1768, BF online
146   'great and hasty demand': *Ibid.*
146   'we are not to wonder': *Ibid.*
147   'great honor': *Ibid.*
147   'If your health and': *Ibid.*
147   Wales delayed by weather: Wales 1770, p.101
148   'most romantic': 19 July 1768, *Ibid.*, p.105
148   'Our situation must': 21 July 1768, *Ibid.*
148   anchored Fort Prince of Wales: 10 August 1768, *Ibid.*, p.116
148   'bear the name of trees': 16 August 1768, *Ibid.*, p.119
148   observatories on bastions of the fort: 19 and 20 August 1768, *Ibid.*
148   'either to speak': 10 August 1768, *Ibid.*, p.118
148   'preferring a Voyage': 18 December 1767, CMRS, vol. v, f.228
148   astronomer to California: Nunis 1982, p.47ff
148   possibility of Pingré to Mexico: 23 August 1768, PV Académie 1768, f.189
149   'sole purpose of this' and following quote: 27 April 1768, Instructions to Spanish Astronomers, reprinted in Nunis 1982, p.114
149   publication of Chappe's book: 31 August 1768, PV Académie 1768, f.202
149   Wales's ship sailed back: 31 August 1768, Wales 1770, p.120
149   'merit of approval': 31 August 1768, PV Académie 1768, f.205
150   'malicious misrepresentation': Catherine II 1772, p.iii
150   Catherine's book *Antidote*: Catherine II 1772. The book was also published in French in 1770. See also Levitt 1998, pp.49–63
150   'I observe you take great': Catherine II 1772, p.34
150   'Admirable genius!': *Ibid.*, p.44
150   'Go! Mons. Chappe': *Ibid.*, pp.7–8
150   fast as 'lightning': *Ibid.*, p.14
150   'No nation . . . has': *Ibid.*, p.8

150 'a new nation': Chappe 1770, p.335
151 Delisle's last days: Anon., 'Éloge de M. de l'Isle', *Histoire & Mémoires*, 1768, p.182
151 Chappe's team for California: Chappe 1778, pp.1-2
151 one 'intervening cloud': *Ibid.*, p.2
151 'make amends to the': *Ibid.*, p.3
151 first snow at Hudson Bay: 8-12 September 1768, Wales 1770, p.120
151 APS offered to travel to Hudson Bay: 20 September 1768, Minutes APS, *Proceedings APS*, 1885, vol. 22, p.18
151 Assembly agreed to purchase telescope: 18 October 1768, *Ibid.*, p.19
151 Assembly later granted £100: Hindle 1964, pp.44-47
151 observatory at State House: 17 February 1769, Minutes APS, *Proceedings APS*, 1885, vol. 22, p.31
151 'at a loss how to furnish', Smith *et al.*, *Transactions APS*, vol. 1, 1769-71, p.9
152 Penn organising telescope in London: *Ibid.*, p.10
152 'an excellent Time-Piece': *Ibid.*, pp.11-12
152 'A particular Account of': advertisement for the *New-England Almanack*, 12 September 1768, *Boston Chronicle*
152 newspapers in America: see also Hindle 1956, p.156
152 transit through 'spy-glass': 27 May 1769, *Providence Gazette*
153 'extreamely important to': John Winthrop to James Bowdoin, 18 January 1769, Bowdoin and Temple Papers, Collections of the Massachusetts Historical Society, Sixth Series, Boston, MHS, 1897, vol. ix, p.116
153 '8 companies to the': *Ibid.*
153 'great pity to lose': *Ibid.*, p.117
153 'without any great expence': *Ibid.*, p.118
153 'principal object': Winthrop 1769b, p.14
153 APS Transit Committee: 3 and 7 February 1769, Minutes APS, *Proceedings APS*, 1885, vol. 22, pp.29-31
153 Rittenhouse observatory: Smith *et al.*, *Transactions APS*, 1769-71a, vol. 1, p.13
153 'least as far westward as': 7 February 1769, Minutes APS, *Proceedings APS*, 1885, vol. 22, p.31
153 'most of the civilized': *Ibid.*, p.30
154 'reputation of their': *Ibid.*
154 'unauthorized' to grant: James Bowdoin to Thomas Gage, 1 March 1769, Bowdoin and Temple Papers, Collections of the Massachusetts Historical Society, Sixth Series, Boston, MHS, 1897, vol. ix, p.130
154 wealthy merchant: this was Joseph Brown. West, *Transactions APS*, 1769-71, vol. 1, p.97
154 gentleman from Newbury: this was Tristam Dalton. Williams, *Transactions APS*, 1786, vol. 2, p.246
154 Surveyor General: this was John Leeds. Leeds, *Phil Trans*, 1769, vol. 59, p.444

13 Racing to the Four Corners of the Globe
155 Lowitz's entourage: Pfrepper and Pfrepper 2009, p.110
155 three-foot-high snow: Fredrik Mallet to Johan Henrik Lidén, 9 September 1769, Heyman 1939, p.284

155 cold and no warm food: Fredrik Mallet to Johan Henrik Lidén, 8 March 1769, *Ibid.*, p.276
155 Mallet regretted being an astronomer: *Ibid.*
155 'ugly Lapp women': Fredrik Mallet to Johan Henrik Lidén, 24 April 1769, *Ibid.*, p.278
155 'There is hardly a day': Planman to Wargentin, 4 April 1769, MS Wargentin, KVA, Center for History of Science
155 'anxious and worried': *Ibid.*
155 loaded musket': Planman to Wargentin, 10 April 1769, MS Wargentin, KVA, Center for History of Science
155 Pingré left Paris: Woolf 1959, p.160; Pingré, *Histoire & Mémoires*, 1770, p.488; Armitage 1953, p.57
155 'ice almost half': Wales 1770, p.124
156 brandy froze: 6 November 1768, Wales and Dymond, *Phil Trans*, 1770, vol. 60, p.144
156 cold stopped clocks: Wales and Dymond, *Phil Trans*, 1769, vol. 59, p.469
156 Maskleyne mounting new expedition: 22 September 1768 and 15 December 1768, CMRS, vol. v, ff.332, 343ff
156 Maskelyne suggesting northern locations: 17 November 1767, 22 September 1768, 15 December 1768, CMRS, vol. v, ff.176, 332, 343 and 16 February 1769, CMRS, vol. vi, f.14
156 Maskelyne and Spitsbergen: 16 February and 2 March 1769, CMRS, vol.vi, ff.14, 16
156 'application' to Admiralty: 15 December 1768, CMRS, vol. v, f.343
156 'for carrying the Observers': *Ibid.*
156 return Mason and Dixon: Mason and Dixon had left America in mid-September 1768 and attended their first meeting at the Royal Society on 10 November. 10 November 1768, CMRS, vol. v, f.334
156 Mason was reluctant: 15 December 1768, CMRS, vol. v, f.344
156 William Bayley as replacement: *Ibid.*
157 Le Gentil to Manila: Le Gentil (Ebeling) 1781, p.46
157 Le Gentil ordered to Pondicherry: *Ibid.*, pp.56-57
157 'obstacles' in his way: *Ibid.*, p.58
157 French governor in Pondicherry: *Ibid.*, p.116
159 Le Gentil left Manila: *Ibid.*, pp.60-61
159 smooth as a 'lake': *Ibid.*, p.69
159 the sea 'roared': *Ibid.*, p.71
159 'abandoning' the ship: Sawyer Hogg 1951, p.91; Le Gentil (Ebeling) 1781, p.72
159 'I took over here': *Ibid.*
159 only a few storms: Le Gentil (Ebeling) 1781, pp.95-96, 108ff
159 'a more fortunate voyage': *Ibid.*, p.114.
159 Le Gentil's arrival in Pondicherry: *Ibid.*, p.115. Le Gentil left Brest on 26 March 1760 and arrived in Pondicherry on 27 March 1768.
159 search for observatory location: *Ibid.*, p.116
159 Le Gentil's observatory: *Ibid.*, pp.116-117; and Sawyer Hogg 1951, p.128
159 I enjoyed at Pondicherry: Sawyer Hogg 1951, p.127
160 'I was more in touch': *Ibid.*, p.128

160 60,000 pounds of gunpowder: Le Gentil (Ebeling) 1781, p.117

160 Le Gentil cleaning his instruments: Sawyer Hogg 1951, p.128

160 achromatic telescope from Madras: Le Gentil (Ebeling) 1781, p.118

160 'exceeded all expectations': Ibid, p.117

160 observed lunar eclipse: Ibid., p.117

160 'from ocean to ocean': Le Gentil to Lanux, 1 October 1768, Le Gentil 1779 and 1781, vol. 2, p.792

161 meeting botanist: 8 August 1768, Sajnovics Journal, Littrow 1835, p.116

161 brother of the Danish Royal Astronomer: 2 July 1768, Ibid., p.103

161 Peder Horrebow: 18 August 1769, Sajnovics Journal, Littrow 1835, p.153; see also Aspaas, pp.13–14

161 hot chocolate: 3 July 1768, Ibid.

161 'most magnificent': 5 July 1768, Ibid., p.104

161 strawberries and cream: Ibid.

161 'bad wine soup': 6 July 1768, Ibid.

162 dangerous mountain: 4 and 9 July 1768, Ibid., pp.104–106

162 an axle broke: 9 July 1768, Ibid., p.106

162 roads ended: 10 July 1768, Ibid.

162 bought new carts: 13–18 July 1768, Ibid., p.108

162 'half the population': 18 July 1768, Ibid.

162 saw snow-capped mountains: 24 July 1768, Ibid., p.112

162 'bad' meals: no date, Ibid., p.102

162 local priests more 'hospitable': no date, Ibid.

162 went to bed hungry: 24 July 1768, Ibid., p.112

162 axles broke regularly: 18, 19, and 26 July 1768, Ibid., pp.109, 113

162 rotten bridges: 22 July and 24 July 1768, Ibid., pp.110, 112

162 danger of deep precipices: 26 July 1768, Ibid., pp.112–113

162 arrival in Trondheim: 30 July 1768, Ibid., p.114

163 whales next to ship: 25 August 1768, Ibid., p.118

163 waves into their cabin: 11 October 1768, Ibid., p.125

163 sea would 'bury' them: Ibid.

163 sailors increasingly happier: Ibid.

163 arrival Vardø: Ibid.

163 location of observatory: 12 October 1768, Ibid., p.126

163 building materials from mainland:14 October 1768, Ibid., p.127

164 lazy carpenters: 13 November 1768, Ibid., p.130

164 observatory completed: 14 December 1768, Ibid., p.132

164 hampered by bad weather: 20 November and 23 December 1768, 2 February 1769, 18 March 1769, Ibid., pp.131, 133–135

164 freezing weather: October and November 1768, Sajnovics Journal, Littrow 1835, p.128

164 Hell feared for their cabin: 20 November 1768, Sajnovics Journal, Littrow 1835, p.131

164 snow had reached the roofs: 1 December 1768, Ibid., p.132

164 'dry, unpalatable Norwegian': 2 November 1768, Ibid., p.129

164 Northern Lights: 20 November 1768, 5 December 1768, and 27 May 1769, Ibid., pp.131, 132, 137

164 shoot birds: 1 December 1768, Ibid., p.131

164 shoot seals: 1 April 1769, Ibid., p.135

165 flying fish: 27 September 1768, Banks Journal

165   Green taught Cook and officers: Aughton 2003, p.26
165   'Indefatigable in making': 23 August 1770, Cook Journal
165   'ducking' ceremony: 25 October 1768, Banks Journal
165   *Endeavour* in Rio: 13 November–8 December 1768, Cook Journal and Banks Journal; Aughton 2003, p.29ff
165   Cook to viceroy: 14 November 1768, Cook Journal
165   Portuguese soldiers rowing around *Endeavour: Ibid.*
166   'certainly did not believe': *Ibid.*
166   'an invented story': 14 November 1768, *Ibid.*
166   'North Star Passing': 14 November 1768, *Ibid.*
166   cleaning and repair regime: 6 November 1768 (and following entries), *Ibid.*
166   *Endeavour* 'heeled down' and following quote: Daniel Solander to Lord Morton, 1 December 1768, Duyker and Tingbrand 1995, p.278
166   'I am a Gentleman': Joseph Banks to Viceroy of Rio de Janeiro, 17 November 1768, BL Add 34744, f.41
166   'French man laying': Joseph Banks to William Philip Perrin, 1 December 1768, Chambers 2000, p.7
166   Cook's and Banks's letters to viceroy: Cook's letters are mentioned in his journal; for Banks's letters and the Viceroy's replies see BL Add 34744, f.41ff
166   'cursd, swore, ravd' and following quote: Joseph Banks to William Philip Perrin, 1 December 1768, Chambers 2000, p.7
166   hot chocolate at Christmas: 25 December 1768, Sajnovics Journal, Littrow 1835, p.133
167   'all hands g[o]t abominably': 25 December 1768, Banks Journal
167   hammocks bashed: 6 January 1769, *Ibid.*
167   ill-fated plant-collecting expedition: 16 and 17 January 1769, Banks Journal and Cook Journal
167   arrival Tahiti: 13 April 1769, Cook Journal
168   'the truest picture of': 13 April 1769, Banks Journal
168   'for the Defence': 18 April 1769, Robert Molyneux Journal, Beaglehole 1999, vol. 1, p.551
168   description of Fort Venus: drawing of 'Venus Fort' by Sydney Parkinson, Parkinson 1784, plate iv; see also 30 April 1769, Banks Journal; 1 May 1769, Cook Journal; Cook, *Phil Trans*, 1771, vol. 61, p.397ff
169   'lusty' females: 28 April 1769, Banks Journal
169   'their politeness to': 14 April 1769, *Ibid.*
169   observatory brought to Fort Venus: 28 April 1769, Robert Molyneux Journal, Beaglehole 1999, vol. 1, p.551
169   Bayley built observatory at North Cape: Bayley, *Phil Trans*, 1769, vol. 59, p.262
169   Hell almost died: 28 April 1769, Sajnovics Journal, Littrow 1835, p.136
169   the instruments on shore: 1 May 1769, Cook Journal
169   stolen quadrant and retrieval: description based on 2 May 1769, Cook Journal, Banks Journal and Sydney Parkinson Journal
169   'every thing that was': 25 April 1769, Banks Journal
169   'prodiges expert': 14 April 1769, Cook Journal
169   'it immediately becomes': 25 April 1769, Banks Journal

169 'with all imaginable': 13 April 1769, Cook Journal
170 'Principle people': 2 May 1769, *Ibid.*
170 Green, Banks and midshipman and following description: 2 May 1769, Banks Journal and Cook Journal
171 Chappe's departure: Chappe 1778, p.1
171 passage twice as long: *Ibid.*, p.3
171 problems with Spanish authorities: *Ibid.*, p.4ff
171 'which a thousand times': *Ibid.*, p.4
171 'If we were retarded ever' and following quotes: *Ibid.*, p.7
171 'instantly fit out': *Ibid.*, p.8
172 met Doz and Medina in Cadiz: *Ibid.*, p.8ff
172 'little nut-shell': *Ibid.*, p.9
172 'frailty of the vessel': *Ibid.*, p.8
172 'the finest ship': *Ibid.*
172 'a transport of joy': *Ibid.*, pp.8–9
172 calculations too 'tedious': *Ibid.*, p.12
172 'tiresome and uniform': *Ibid.*
172 'a thousand times': *Ibid.*, p.14
172 seventy-seven days to cross the Atlantic: *Ibid.*, p.12
172 'in greatest anxiety': *Ibid.*, p.19
173 instruments carried by mules: *Ibid.*, p.23
173 murderous 'banditti': *Ibid.*, p.46
173 roads were 'frightful': *Ibid.*, p.30
173 heat 'excessive': *Ibid.*
173 'especially for a Frenchman': *Ibid.*, p.24
173 'a most frightful neck': *Ibid.*, p.34
173 'no very pleasing figures': *Ibid.*, p.35
173 Chappe had to wait: *Ibid.*, p.47
173 Chappe's arrival San Blas: *Ibid.*, p.52
173 'to make a tent': *Ibid.*, p.54
173 'began to despair': *Ibid.*, p.55
173 'most cruel disappointment': *Ibid.*, p.56
173 'I little cared whether': *Ibid.*, p.57
174 quarrel Chappe with Doz and Medina: *Ibid.*, p.57ff
174 'rather lose a poor': *Ibid.*, p.58
174 Doz and Medina blaming Chappe: *Ibid.*, p.59
174 'to keep it dry': *Ibid.*, p.61
174 'horrid roaring': *Ibid.*
174 'all their might': *Ibid.*

14 Day of Transit, 3 June 1769

175 Bayley at the North Cape: Bayley, *Phil Trans*, 1769, vol. 59, p.262
175 Dixon in Hammerfest: Dixon, *Phil Trans*, 1769, vol. 59, p.253
175 Dixon's arrival Hammerfest: Acerbi 1802, vol. 2, p.117
175 Astronomers in Germany: Mayer, Andreas, *Phil Trans*, 1769, vol. 59, pp.284–285; Lalande, *Phil Trans*, 1769b, vol. 59, p.376; 14/25 August and 9/20 November 1769, Protocols, vol. 2, pp.69, 713; Moutchnik 2006, p.207; *Göttingische Anzeigen von Gelehrten Sachen*, 1769, vol. 1, p.665; 25 January 1770, JBRS, vol. 27, f.286
175 Dutch observers in Leiden: Johan Henrik Lidén to Fredrik Mallet, 10 July 1769, Heyman 1938, p.280

175   Planman in Kajana: Planman to Wargentin, 10 April 1769, MS Wargentin, KVA, Center for History of Science

175   Mallet in Pello: KVA, Verification 162; Nordenmark 1946, p.75

175   Krafft in Orenburg: 10/21 April 1769, Protocols, vol. 2, p.680

176   Euler in Orsk: Euler 1769, p.8

176   Lowitz racing against thaw: 12/23 April 1769, Lowitz 1770, pp.20–21

176   Lowitz left behind entourage: 8/19 May 1769, Ibid., p.21

176   Mayer in observatory: 8/18 May 1769, Protocols, vol. 2, p.682; Moutchnik 2006, p.201

176   Christian Mayer's journey to Russia: Moutchnik 2006, p.186ff

176   'a bit strange': Johann Albrecht Euler to Jean Henri Samuel Formey, 24 March/4 April 1769, in Moutchnik 2006, p.194

176   Mallet, Pictet and Rumovsky on Kola peninsula: 27 April/8 May and 20/31 May and 26 May/6 June 1769, Protocols, vol. 2, pp.681, 686, 687

176   Islenieff in Yakutsk: 4/15 May 1769, Protocols, vol. 2, p.682

176   observer in Mexico: this was Joaquin Velázquez de León who was a self-taught astronomer. Nunis 1982, p.71ff

176   'Surveyor-General of Lands': Holland, *Phil Trans*, 1769, vol. 59, pp.247–252

176   Winthrop in Cambridge, MA: Winthrop, *Phil Trans*, 1769a, vol. 59, pp.351–358

176   'after much Delay': Benjamin Franklin to Winthrop, 11 March 1769, BF online

176   telescopes for APS: Joseph Shippen to Edward Shippen, 29 May 1769, Papers of Edward Shippen, DLC

176   Chappe's arrival: Chappe 1778, p.64ff

177   'epidemical distemper': *Ibid.*, p.64

177   'not stir from San-Joseph': *Ibid.*

177   'makes us all not': 31 May 1769, Banks Journal

177   'for fear that we': 30 May 1769, Cook Journal

178   'very buisy': 30 May 1769, *Ibid.*

178   first team left: 1 June 1769, Cook Journal and Banks Journal

178   second team left: 2 June 1769, Cook Journal

178   'all Hands anxious': 2 June 1769, Robert Molyneux, Remarks in Port Royal Bay in King George the thirds Island, Beaglehole 1999, vol. 1, p.559

178   'prov'd as favourable': 3 June 1769, Cook Journal

178   'might disturb the Observation': 3 June 1769, Robert Molyneux, Remarks in Port Royal Bay in King George the thirds Island, Beaglehole 1999, vol. 1, p.559

178   'first visible appearance of': Cook, *Phil Trans*, 1771, vol. 61, p.410

178   entry for Green: *Ibid.*

178   entry for Solander: *Ibid.*, p.412

178   'wavering haze': *Ibid.*

178   some 'undulating': *Ibid.*, p.410

178   'very difficult to judge': *Ibid.*, p.411

178   119°F

178   'intolerable': *Ibid.*, p.411

178   twelve seconds apart: *Ibid.*, p.410

178   timing 'a little doubtful': *Ibid.*, p.411

180 entertainment in Tahiti during transit day: 3 June 1769, Banks Journal and 4 June 1769, Cook Journal

180 'I had still time enough' and following quotes: Chappe 1778, p.62

180 Chappe's observatory: 'On the Observations made by Abbé Chappe in California', transcribed and translated in Nunis 1982, p.105ff; see also Chappe 1778, p.63

180 instruments 'as they were': Chappe 1778, p.63

180 'groans' of the stricken inhabitants: *Ibid.*, p.65

180 'cared for nothing else': *Ibid.*

181 'the weather . . . favoured me': *Ibid.*, p.63

181 'might at every moment': 'Astronomical Observations Made in the Village of San José in California', transcribed and translated in Nunis 1982, p.99

181 fixing instruments: 'On the Observations made by Abbé Chappe in California', *Ibid.*, p.105

181 clock protected from dust: 'Astronomical Observations Made in the Village of San José in California', *Ibid.*, p.98

181 tasks of team on transit day: Nunis 1982, p.69

181 'detaching itself with': 'On the Observations made by Abbé Chappe in California', *Ibid.*, p.106

181 Chappe's times: Nunis 1982, p.68

181 'I had an opportunity': Chappe 1778, p.63

181 snow storm Vardø: 1 May 1769, Sajnovics Journal, Littrow 1835, p.136

182 news from Hammerfest and North Cape: 12 May 1769, *Ibid.*, p.137

182 news from Kildin: 14 May 1769, *Ibid.;* for Ochtenski in Kildin, 12/23 May 1769, Protocols, vol. 2, p.684

183 'great observation': 2 June 1769, *Ibid.*, p.138

183 Vardø day of transit: 3 June 1769, *Ibid.*, pp.138–139

183 fog at Arctic Circle: Aspaas 2008, p.16

183 weather 3 June 1769: 3 June 1769, Sajnovics Journal, Littrow 1835, pp.138–139

183 'with the special grace': *Ibid.*, p.139

183 'Unbelievable! . . . but nevertheless': *Ibid.*

184 'Te Deum laudamus': *Ibid.*

184 'wish me luck': Le Gentil (Ebeling) 1781, p.118

184 Pondicherry evening before transit: *Ibid.*

184 'With my soul content': Sawyer Hogg 1951, p.127

184 'moaning' of sandbanks: *Ibid.*, p.131

184 'From that moment': *Ibid.*

185 'a second curtain': *Ibid.*

185 a 'light whiteness': *Ibid.*

185 sun after transit: Le Gentil (Ebeling) 1781, p.119

185 'I had difficulty in': Sawyer Hogg 1951, p.132

185 clouds appeared to cause 'chargrin': Le Gentil (Ebeling) 1781, p.119

185 'spectator of a fatal cloud': Sawyer Hogg 1951, p.131

185 'purity of atmosphere': Smith *et al.*, *Transactions APS*, 1769–71a, vol. 1, p.23

185 audience at Norriton: *Ibid.*, p.25

185 Rittenhouse fainted: Rush 1796, pp.12–13

185   'without any rest': 23 May–3 June 1769, Catherine's Court Journals 1769, Sobolevskii 1853–67, pp.96–97. Catherine observed the transit in Bronnaya near her summer residence Oranienburg.

186   King George III observed transit: Demainbray, Stephen, 'Observations on the Transit of Venus', The King George III Museum Collection, King's College London, K/MUS 1/1 1768–69

186   Peking observation: Cipolla, *Phil Trans*, 1774, vol. 64, pp.31–45

186   Mason in Ireland: Mason, *Phil Trans*, 1770, vol. 60, pp.454–496

186   Madras observation: Le Gentil (Ebeling) 1781, p.119

186   Jakarta observation: Mohr, *Phil Trans*, 1771, vol.61, pp.433–436; Zuidervaart and Van Gent 2004, pp.15-16

186   Banks and Solander at Mohr's observatory: Zuidervaart and Van Gent 2004, p.18

186   two amateurs trained by Le Gentil: Le Gentil (Ebeling) 1781, p.119

187   transit projected on wall: Martin 1773, p.28

187   '*Artificial* Transit': *Ibid.*, p.23; see also Bernoulli 1771, pp.73–74

187   'most of the Inhabitants': 10 June 1769, *Providence Gazette*

187   'were totally deprived': 10 July 1769, *New York Gazette*

187   Crowds watching transit: for example, in Louis XV's Château de la Muette near Paris, in Vardø, in Newbury, Massachusetts and in Norriton, Pennsylvania. Smith 1954, p.448; 3 June 1769, Sajnovics Journal, Littrow 1835, p.139; Williams, *Transactions APS*, 1786, vol. 2, p.248; Smith *et al.*, *Transactions APS*,1769–71, vol. 1, p.25

188   'earthly Venus': Johan Henrik Lidén, 3 June 1769, in Hulshoff Pol 1958, p.144.

188   'looked like she would have allowed some immersion': *Ibid.*

188   some 'young Bloods': 3–6 June 1769, *London Chronicle*

188   'made a *Transit* into': *Ibid.*

15 After the Transit

189   Chappe's illness: Chappe 1778, pp.63–73; Nunis 1982, p.93ff

189   'a scene of horror': Chappe 1778, p.65

189   'either dying or hastening': *Ibid.*, p.66

189   Chappe observed lunar eclipse: *Ibid.*, p.68

189   observations despite illness: *Ibid.*

189   transit notes in box: Chappe, *Phil Trans*, 1770, vol. 60, p.551

189   'mere desert': Chappe 1778, p.69

189   attacked by insects: Nunis 1782, p.79

190   Vicente de Doz returned: Masserano, Phil Trans, 1770, vol. 60, pp.549–550; Nunis 1982, p.124ff

190   engineer and painter survived: *Ibid.*, p.71

190   'true philosopher': Chappe 1778, p.70

190   Chappe's notes to Paris: Nunis 1982, p.75. The engineer Pauly brought Chappe's manuscripts back to Paris in December 1770

190   Chappe's funeral: Chappe 1778, p.70

190   Le Gentil after transit: Sawyer Hogg 1951, p.132; Le Gentil (Ebeling) 1781, p.119

190   sky had been 'cruel': Le Gentil (Ebeling) 1781, p.119

191   fifty-one copies dispatched: 29/18 September 1769, Protocols, vol. 2, p.703

191   observations Christian Mayer: Proctor 1882, p.61

191 'wave-like' edge of Venus: Krafft 1769, p.19

191 rain on Kola peninsula: this was Pictet. Euler to Lowitz, 13/24 July 1769, transcribed in Pfrepper and Pfrepper 2009, p.113; see also *Göttingische Anzeigen von Gelehrten Sachen*, 1769, vol. 1, p.1,143

191 Rumovsky's assistant died on Kildin: 14 May 1769, Sajnovics Journal, Littrow 1835, p.137

191 Rumovsky in Kola: Euler to Lowitz, 13/24 July 1769, transcribed in Pfrepper and Pfrepper 2009, p.113; see also *Göttingische Anzeigen von Gelehrten Sachen*, 1769, vol. 1, p.1,144

191 astronomer in Göttingen: this was Gotthilf Kästner, *Göttingische Anzeigen von Gelehrten Sachen*, 1769, vol. 1, p.665; and 14/25 August 1769, Protocols, vol. 2, p.697

191 'nothing' in Denmark: 5/16 October 1796, Protocols, vol. 2, p.707

191 'one of the most beautiful': Wargentin, KVA Abhandlungen, 1769, p.148

191 stayed up all night in Sweden: Prosperin, KVA Abhandlungen, 1769, p.158

191 French observations: Lalande, *Phil Trans*, 1769b, vol. 59, p.374ff

192 'noise and confusion': Smith 1954, p.448

192 'I was precisely in': Lalande, *Phil Trans*, 1769b, vol. 59, p.375

192 British struggled: for example, Hornsby, *Phil Trans*, 1769, vol. 59, p.172; Wollaston, *Phil Trans*, 1769, vol. 59, p.408; Harris, *Phil Trans*, 1769, vol. 59, p.425; Bradley, *Transactions APS*, 1769–71, vol. 1, p.115

192 observations in Greenwich: Maskelyne, *Phil Trans*, 1768b, vol. 58, p.361

192 'a degree of correctness': Delambre, 'Life and Works of Dr Maskelyne, 1813', RGO 4/226, p.22

192 'greater than I expected': Maskelyne, *Phil Trans*, 1768b, vol. 58, p.362; Hornsby thought the same: Hornsby, *Phil Trans*, 1769, vol. 59, p.176

192 billowing smoke in London: Horsfall, *Phil Trans*, 1769, vol. 59, p.170; Canton, *Phil Trans*, 1769, vol. 59, p.192

192 'begging the inhabitants': Wilson, *Phil Trans*, 1769, vol. 59, p.334

192 Venus's edge 'bubbled': Prosperin, KVA Abhandlungen, 1769, p.156

192 'like the waves of a': Ferner, *Phil Trans*, 1769, vol. 59, p.404

192 'an apple connected': Prosperin, KVA Abhandlungen, 1769, p.156

192 the neck of a Florence': Bevis, *Phil Trans*, 1769, vol. 59, p.190

192 'pointed truffle': Ferner, *Phil Trans*, 1769, vol. 59, p.405

192 'small shadow': Gissler, KVA Abhandlungen, 1769, p.226

192 'dark thread': Gadolin, KVA Abhandlungen, 1769, p.173

192 'Venus's circular figure': Maskelyne, *Phil Trans*, 1768b, vol. 58, p.358

192 Venus was 'misshapen': Gissler, KVA Abhandlungen, 1769, p.225

192 Venus was 'ill defined': Harris, *Phil Trans*, 1769, vol. 59, p.425

192 luminous ring: Krafft 1769, p.38; Maskelyne, *Phil Trans*, 1768b, vol. 58, p.359

192 'tremulous vapours': Schenmark, KVA Abhandlungen, 1769, p.223

192 observers in Stockholm: Wargentin, KVA Abhandlungen, 1769, pp.148–154

192 observers in Uppsala: Prosperin, KVA Abhandlungen, 1769, pp.156–158

192 observers in Paris: Lalande, *Phil Trans*, 1769b, vol. 59, pp.374–377

192   observers in Greenwich: Maskelyne, *Phil Trans*, 1768b, vol. 58, p.358

192   observers in St Petersburg: Proctor 1882, p.61

192   observers in Orenburg: Krafft 1769, p.19ff

193   Mallet's report from Pello: 5 July 1769, KVA Protocols, p.875

193   Mallet's day of transit: Wargentin, KVA Abhandlungen, 1769, p.147; Mallet, KVA Abhandlungen, 1769, p.218ff; Fredrik Mallet to Johan Henrik Lidén, 9 September 1769, Nordenmark 1946, pp.76–77

193   'miserable night': Fredrik Mallet to Johan Henrik Lidén, 9 September 1769, Nordenmark 1946, p.77

193   'fallen out' with Venus: *Ibid.*

193   'Never . . . have my': Planman to Wargentin, 9 June 1769, MS Wargentin, KVA, Center for History of Science

194   'tears' in his eyes: *Ibid.*

194   Planman's day of transit: Planman, KVA Abhandlungen, 1769, p.212; Planman to Wargentin, 9 June 1769, MS Wargentin, KVA, Center for History of Science

194   Stockholm and London reports in Paris: 28 June 1769, PV Académie 1769, f.235

194   Planman reports in Paris 9 August 1769, PV Académie 1769, f.298

194   Wargentin's report in St Petersburg: 30 June/11 July 1769, Protocols, vol. 2, p.693

194   North American reports in London: Maskelyne to Thomas Penn, 2 August 1769, read at the APS on 15 December 1769, Minutes of Meetings, APS, *Proceedings APS*, vol. 22, 1885, p.46

194   Russian reports in Kajana: 5/16 October 1769, Protocols, vol. 2, p.707

194   Russian observations in Germany: *Göttingische Anzeigen von Gelehrten Sachen*, 1769, vol. 1, p.143

194   North American results published in *Phil Trans*: Wright, *Phil Trans*, 1769, vol. 59, p.273ff; Smith *et al.*, *Phil Trans*, 1769, vol. 59, p.289ff

195   Lalande reports about missionary and Pingré: Lalande, *Phil Trans*, 1769b, vol. 59, p.376

195   Pingré in Haiti: Pingré, *Histoire & Mémoires*, 1770, p.503

195   observations Dixon and Bayley: 15 November 1769, JBRS, vol. 27, f.244; Bayley, *Phil Trans*, 1769, vol. 59; Dixon, *Phil Trans*, 1769, vol. 59

195   Wales returned: 9 November 1769, CMRS, vol. vi, f.55. Wales had returned to London in October.

195   successful observations Wales: Wales and Dymond, *Phil Trans*, 1769, vol. 59, p.480ff

195   Lalande and Hell: Sarton 1944, p.100; Woolf 1959, pp.178–179

195   Hell and anti-Jesuit sentiment: Sarton 1944, p.102

195   Hell in Copenhagen and rumours: Hell arrived in Copenhagen on 17 October 1769. Sajnovics Journal, Littrow 1835, p.157; Moutchnik 2006, p.234; Woolf 1959, p.178; Hamel *et al.* 2010, p.190

195   120 copies of Vardø observations: Sajnovics Journal, Littrow 1835, p.158

195   accusations Hell: Sarton 1944, p.104; Woolf 1959, pp.178ff

195   Chappe's engineer arrived in Paris: Chappe, *Phil Trans*, 1770, vol. 60, pp.551–552

196   Botany Bay: 6 May 1770, Cook Journal

196   Banks collected so many plants: 3 May 1770, Banks Journal

197   *Endeavour* struck reef: 10–12 June 1770, Cook Journal and Banks Journal
197   'fear of Death now': 11 June 1770, Banks Journal
197   'rising all most perpendicular': 18 August 1770, Cook Journal,
197   Green continues observations: Beaglehole 1999, vol. 1, p. cxxxiv
197   crushed 'to peices': 18 August 1770, Cook Journal
197   hull one-eighth of an inch: 14 November 1770, Banks Journal
197   Jakarta: 12 October–26 December 1770, Cook Journal and Banks Journal
198   Green died: 29 January 1771, Cook Journal
198   'In a fit of phrensy': 27 July 1771, *General Evening Post*
198   'he took no care': 29 January 1771, Cook Journal
198   Le Gentil's heirs spread rumours: Le Gentil (Ebeling) 1781, p.127
198   natural history collections in Mauritius: *Ibid.*, p.126
198   attacks of dysentery: *Ibid.*, p.123
198   weakened by illness: Le Gentil to Lanux, 27 April 1770, Le Gentil 1779 and 1781, vol. 2, pp.796–797
198   lost belief to see France again: Le Gentil (Ebeling) 1781, p.127
198   'had become unbearable': Sawyer Hogg 1951, p.174
199   Le Gentil boarded ship to Europe: Le Gentil (Ebeling) 1781, p.129
199   'than I had ever seen before': *Ibid.*
199   lay down and waited to die: Le Gentil to Lanux, 2 February 1772, Le Gentil 1779 and 1781, vol. 2, p.800
199   'tired of the sea': *Ibid.*, p.806
199   heirs declared him dead: Le Gentil (Ebeling) 1781, pp.140–141; Sawyer Hogg 1951, p.177

Epilogue: A New Dawn

200   'differ more from one': Cook, *Phil Trans*, 1771, vol. 61, p.406
200   'publish to the world': 17 September 1773, Beaglehole 1999, vol. 2, p.238
201   final results at RS: read on 19 December 1771, Hornsby, *Phil Trans*, 1771, vol. 61, pp.574–579
201   parallax was 8"78: *Ibid.*, p.579
201   93,726,900 miles: *Ibid.*
201   'uncertainty . . . entirely removed' and following quotes: *Ibid.* p.574
201   calculations ranged from 8"43 to 8"80: Nunis 1982, p.69
201   Kepler's and Halley's predictions: Woolley 1969, p.29
203   Hell's planned publication: Hansen and Aspaas 2005, pp.8–9; Sarton 1944, p.101
203   'advantage of the empire': Instructions for Pallas, in Bacmeister 1772, vol. 1, p.89; see also Wendland 1992, vol. 1, p.91
203   new map of Russia: 15/27 April 1770, Protocols, vol 2, p.740
204   Pallas meeting Lowitz: Wendland 1992, vol. 1, p.96; Lowitz and canals: Pfrepper and Pfrepper 2004, p.171
204   Lowitz murdered: Pfrepper and Pfrepper, 2004, p.172
204   results of Pallas's expedition: Wendland 1992, vol.1, p.140ff; Vucinich 1984, p.25
204   *Endeavour* returned with 30,000 specimens: Carter 1988, p.95
204   'vast book of information': Joseph Banks to George Yonge, 15 May 1787, Chambers 2000, p.89

204   Banks as promoter of the colonisation: Carter 1988, p.216
204   Banks and First Fleet: Wulf 2008, p.211; Gascoigne 1994, p.203
204   Banks and plants for colonial progress: Wulf 2008, p.208ff
205   'The science of two Nations': Joseph Banks to Jaques Julien Houttou
      de La Billardière, 9 June 1796, Chambers 2000, p.171
205   prize for obtaining saltpetre: McClellan 1985, pp.175, 338

Index

Académie des Sciences, Paris
see French Academy of
Sciences
Adams, John, US President
81
Adolph Frederick, King of
Sweden 134, 135, 136
Aepinus, Franz 46, 47–8, 58,
75, 76, 113, 115, 185
Alembert, Jean Le Rond d'
(with Diderot):
Encyclopédie xxv, 112,
127
American Philosophical
Society (APS) 142–3,
144, 146, 151, 152,
153, 176, 185, 200
Audiffreddi, Giovanni
Battista 98

Banks, Joseph: on the
Endeavour 132, 165,
166, 167; on Tahiti
169, 170, 180n; in
Jakarta 186n; at Botany
Bay 196; and accident
on Great Barrier Reef
197; as president of
Royal Society 204–5

Bavarian Academy of
Sciences, Munich 52,
60, 83
Bayley, William 156, *157*,
169, 175, 182, 195
Bencoolen, Sumatra 15–16,
31, 37, 38, 39, 40, 41,
43, 57, 62–3, 81, 84
Berthoud, Ferdinand 108
Bird, John 146
Borchgrevink, Jens Finne
138n
Boscovich, Fr Roger Joseph
109 *and n*
Boston Chronicle 152
Boston Evening Post 57, 86
Bougainville, Captain
Louis-Antoine de 167n
Buffon, Georges-Louis
Leclerc, comte de 143

Cape Henlopen, Delaware
151
Cape of Good Hope 3, 8,
19–20, 27, 62, 64, 66,
79–80, 89–90, 98, 199
Cape Town 6, *63*, 63–4
Cape Verde islands 27, 32
Carlos III, of Spain 101,

102, 109, 141, 142,
173
Catherine II ('the Great'):
becomes Empress
101–2; and Voltaire
112; interest in science
and astronomy 42n,
102, 111, 112–13; and
Aepinus 46, 113, 115,
185; her plans for
Russia 111, 112–13,
115n, 143; enraged by
Chappe's book 150;
and transit expeditions
115, 116–17, 118,
119–20, 121, 125, 140,
146, 153, 191, 203,
204; views 1769 transit
185–6
Chappe d'Auteroche,
Jean-Baptiste 12, 41,
43; journeys to St
Petersburg 43–6; on to
Siberia 48–51, 60; sets
up Tobolsk observatory
65; and 1761 transit
72–4, 92, 96; at Paris
Observatory 108;
publishes Voyage en
Sibérie 108, 149, 150,
203; and 1769 expedi-
tion to California
124–5, 141, 148–9,
150–51, 155, 171–4,
176–7; observes transit
180–81; illness and
death 189–90; his
results saved 190, 195,
201
Charles XII, of Sweden 52

Charlotte, Queen 140n
Christian VII, of Denmark
101, 137, 138, 139,
140, 195
Clark, William 205
Clement XIV, Pope 203n
Clive, Robert 18, 21, 124
Compagnie des Indes (French
East India Company)
12, 20, 21, 22, 25, 26,
27, 28, 66, 107
Comte d'Argenson (ship)
25–8
Cook, Elizabeth 132
Cook, Captain James: meets
Franklin 146; prepares
for South Sea expedi-
tion 128–9, 130–32;
voyage on Endeavour
143, 156, 165–7; on
Tahiti 167–70, 175;
observations and results
177–8, 179, 180, 200n,
201n, 202; makes
landfall in Australia
196; runs aground
196–7; loses men in
Jakarta 197–8; arrival
in Britain 198, 200
Copenhagen: Academy of
Sciences 195
Copernicus, Nicolaus
xxv–xxvi
Cortés, Hernán 172n
Crabtree, William 17n

Darwin, Charles 205
Delaware 92, 144n, 151
Delisle, Joseph-Nicolas 6–7;
backs Halley's idea of

coordinating transit observations 4–6, 7–8, 10–11, 14, 29, 31, 37, 41, 52, 68, 69, 85; meets Halley 7; his 'mappemonde' 8, 9, 11, 13, 14, 55, 58; and Le Gentil 12, 22; and Chappe 12, 42–3; distributes *Vénus passant sur le Soleil* 58; watches 1761 transit 85; his pupils 95, 103; leaves scientific life 141–2; last appearance at Académie 151; death 150–51

Diderot, Denis 112; *Encyclopédie* (with d'Alembert) xxv, 112, 127

Dixon, Jeremiah 37; sets out for Bencoolen 37–41, 58; arrives at Cape of Good Hope 60, 62–4, 66; observes 1761 transit 79–80; further observations 89–90; reputation restored 91; results questioned 98; and 'Mason–Dixon line' 91–2, 124, 125; accepts North Cape commission 156; views 1769 transit from Hammerfest 175, 182, 195

Dolphin (ship) 167*n*

Doz, Vicente de 172, 173, 180, 190*n*

Dutch East India Company 58, 64

East India Company, British 15–16, 17–18, 19, 31, 34, 37, 38, 122, 125, 186

Edinburgh Magazine 57

Endeavour (ship) 128–9, 130–32, 140, 144, 156, 165–9, 175, 177–8, 180*n*, 186*n*, 196, 196–8, 204

Enlightenment, the xxiv, xxv, 10, 13, 90, 101, 112, 113, 116, 131

Ferguson, James 58 *and n*, 84

Flamsteed, John 35

Franklin, Benjamin xxv, 4, 142, 143, 144, 145–7, 153, 176

French Academy of Sciences (Académie des Sciences) 4, 5, 34, 58, 90, 91, 107, 194, 205; and Le Gentil 12, 21, 22, 89, 107, 110, 199; and Pingré 22, 23, 24, 25, 105; and Chappe 43, 48, 92, 98, 124–5, 141, 148, 149, 150, 151, 190

French East India Company *see* Compagnie de Indes

Galileo Galilei xxvi

George II, of England 13, 17, 20, 34, 57

George III, of England 101, 102, 126, 128, 140 *and n*, 146, 186
Glasgow (1769) 192
Goethe, Johann von 112
gravity-measurement tests 89–90
Green, Charles 129, 132, 146; on expedition to South Sea 165; on Tahiti 169, 170; observation of 1769 transit and results 178, *179*, *180*, 200, 201*n*, 202*n*; and survey of Australian coastline 196, 197; death 198
Greenwich: Royal Observatory 10, 31, 34, 35, 37, 123, 124, 129–30, 140*n*, 156, 192; meridian 33, 35, 95, 127
Gustav, Crown Prince of Sweden 78, 101

Haiti 3, 148*n*, 155, 195
Halley, Edmond xxviii; observes transit of Mercury (1677) 61–2; calls on scientists to unite in observing transit of Venus xxiii–xxiv, xxv, xxvi, xxviii, 5, 10, 11; drawing of transit of Venus *xxix*; astronomical tables 7, 43*n*; method of duration 5, 11, 69–70; names important observation locations 29, 37;

meets Delisle 7; treatise translated 58; predicts solar parallax 201
Hammerfest, Norway 157, 175, 182, 195
Harrison, John 122–3, 129
Hell, Maximilian: expedition to Vardø 137, 138–40, 156, 161–4, *163*, 166–7, 169*n*, 181–4; maps *161*, *182*; rumoured to have failed to view transit 195; results 200, 201; plans book 203 *and n*
Hooke, Robert xxv
Hornsby, Thomas 98*n*, 105, 201
Horrebow, Christian 161*n*
Horrebow, Peder 161 *and n*
Horrocks, Jeremiah xxiv, 17, 97
Hudson Bay 7, 10, 123, 125, 127, 129–30, 141, 146, 148–9, *149*, 151, 155, 175, 195; results 201, 202
Hudson's Bay Company 126, 127–8, 149

Imperial Academy of Sciences, St Petersburg 42, 46; and Chappe 46, 48, 60; and Lomonosov 46–7, 75; and reports of 1761 transit 90, 91, 92; and Rumovsky/Popov battle 97; and Catherine the Great 113–17, 120, 125; and 1769

expedition 119;
Wargentin's report to
136; and Franklin 146;
dispatches reports
efficiently 191; results
calculated by Lexell
200; commissions new
empire map 203*n*; calls
back scientific teams
204

Jakarta 3, 6, 10, 11, 15, 16,
22, 55, 58, 60, 68,
186, 196, 197, *197*
Jefferson, Thomas 145
Jesuits/Society of Jesus 52,
83, 109, 137, 138, 191,
195, 203*n*
Johnson, Samuel: *Dictionary*
xxv
Jupiter xxvi *and n*; satellites
33, 62, 67, 127, 184

Kajana, Finland 6, 41, 55,
56, 65, 77–8, 136, 155,
175, 193–4
Kant, Immanuel xxv
Karl Theodor, Elector
Palatine 83
Kepler, Johannes xxvi *and n*,
201
Kildin island 182, 191
Kloster Berge, Germany:
observatory 93
Kola peninsula 114, 115,
119, 140, 176, 191;
results 201, 202

Lalande, Jérôme 17*n*, 142;
on Delisle 7; collates

1761 transit data 95–6;
and 1769 transit 103
and n, 192, 200; and
C. Mayer 176; and Hell
195
latitudes 32
Le Gentil de la Galaisière,
Guillaume Joseph
Hyacinthe Jean-Baptiste
11–12; attempts to
reach India 12, 19–22,
24, 28–30; in Mauritius
22, 24, 30, 45, 60;
observes 1761 transit
aboard ship 71–2, 87;
returns to Mauritius
87, 88–9; plans to chart
islands 88–9, 106, *106*,
203; prepares for 1769
transit 107; sets off for
Manila 107, 110, 186;
ordered to go to India
110, 126, 157, 159;
maps route *158*; builds
observatory 159–60,
160, 175; sees nothing
184–5, 190; rumours of
death spread by heirs
198, 199; trapped in
Mauritius 198–9;
returns to France 199;
writes *Voyage dans les
mers de l'Inde* 203
Leiden, Netherlands 83, 175,
188
Lewis, Meriwether 205
Lexell, Anders Johan 93*n*,
200
Linnaeus, Carl xxv, 55*n*,
138*n*

Littrow, Carl Ludwig 195*n*
Lomonosov, Mikhail 46–8,
    75–7, 91, 93, 113
longitude 31, 32 *and n*,
    33–6, 108
Longitude Prize 33, 34, 35,
    122, 123 *and n*
Louis XIV, of France 5
Louis XV, of France 5, 11,
    23, 85, 101, 191
Louisa Ulrika, Queen 78,
    101
Lowitz, Georg Moritz 117
    and n, 155, 176, 203–4
Lys, Le (ship) 27, 28

Madagascar 89, 106, 203
Magdeburgische Zeitung 77*n*
Mallet, Fredrik 135–7, 140,
    155, 175, 193, 203
Mallet, Jacques André 117*n*,
    135*n*
Manila, Philippines 3, 107,
    110, 157, 159, 186,
    190
Martin, Benjamin 58, 86,
    187; writes *Venus in
    the Sun* 58; his
    'Artificial Transit' *186*,
    187
Martinique 195
Maryland 91–2
Maskelyne, Nevil: appointed
    observer on St Helena
    17–18, 31; devises
    lunar method of calcu-
    lating longitude 33,
    34–6, 83, 122; lands at
    St Helena 36–7, 61;
    sets up observatory
    61–2; observes 1761
    transit 80–81; conducts
    gravity experiments 89,
    90; sends results to
    Royal Society 91, 93,
    98; and John Harrison
    122–3, 129; made
    Astronomer Royal 123;
    heads Royal Society
    'Transit Committee'
    122; plans for 1769
    transit 123–4, 125–6;
    produces *Nautical
    Almanac* 127 *and n*;
    plans Hudson Bay and
    South Sea expeditions
    127, 128, 129–30, 131,
    132, 148; and Russian
    expedition 136; recom-
    mends Vardø as obser-
    vation site 137–8; and
    George III 140 *and n*;
    involves US astrono-
    mers 146–7, 151–2,
    153, 194; plans expedi-
    tion to North Cape 156
    and n; and inaccuracies
    of British observations
    192; displeased with
    Cook's results 200
Mason, Charles: at
    Greenwich Observatory
    37; sets out for
    Bencoolen 37–41, 58;
    arrives at Cape of
    Good Hope 60, 62–4,
    66; observes 1761
    transit 79–80; further
    observations 89–90;
    reputation restored 91;

results questioned 98; and 'Mason–Dixon line' 91–2, 124, 125; refuses North Cape commission 156; in Ireland for 1769 transit 156*n*, 186
Mauritius 20, 21, 22, 24, 27, 28–9, 30, 45, 60, 66, 88, 106, 107, 198–9
Mayer, Fr Christian 83, 176, 191
Mayer, Tobias 35, 83, 123*n*
Medina, Salvador de 172, 173, 180, 190, 198
Mercury, transits of 6, 12, 24, 43*n*, 61–2, 81
Mignonne (ship) 87, 88
Montesquieu, Charles de Secondat, baron de 112
Morton, Charles 41

Napoleon Bonaparte 36*n*; army 205
New-England Almanack 152
Newton, Sir Isaac 13; *Principia* xxvi, xxviii
Norriton, Pennsylvania 144, 151, 152, 153, 185
North Cape, Norway 123, 140, 156, 157, 169, 175, 182, 185
Novelle Letterarie 85

Ochtenski (Russian astronomer) 182
Oiseau (ship) 87, 88
Orenburg, Russia 119, 175, 191, 192, 203
Orlov, Count Grigory 111

Orlov, Count Vladimir 111, 115
Orsk, Russia 119, 176

Pallas, Peter Simon 203, 204
Paris: Académie des Sciences *see* French Academy of Sciences; Royal Observatory 12, 31, 33
Paul I, of Russia 113
Pello, Lapland 135, 136, 137, 140, 175; results 193
Penn, Thomas 152
Penn, William 152*n*
Pennsylvania 91–2, 152, 194, *see also* Norriton
Peter I ('the Great'), of Russia 42, 111
Peter III, of Russia 101–2
Philadelphia xxv, 143–4, 151, 153, 185
Pictet, Jean-Louis 117*n*
Pingré, Alexandre-Gui 23–4; sent to Rodrigues 12, 22, 23, 24–8, 41, 60; reaches Mauritius 28 *and n*, 66; notes inaccuracy of sea charts 32; reaches Rodrigues 66–7; observes 1761 transit 70 *and n*, 74–5; stranded by British 87–8, 91; and his observations 88, 89, 95, 96, 98; proposes destinations for 1769 expedition 105–6, 110, 124; views transit from Haiti 148*n*, 155, 195; in Paris 200

Planman, Anders: reaches
Kajana 55–6, 60; views
lunar eclipse 65–6;
observes 1761 transit
70 *and n*, 77–8, 134;
'takes the waters' in
Oulu 89, 94; changes
his results 94–5, 96;
obsessively recalculates
data 136, 137; observes
1769 transit 136, 155,
175, 193–4, 200
Plassey, Battle of (1757) 21
Pondicherry, India 3, 7, 10,
12, 20, 21, 22, 29, 30,
41, 42, 57, 81, 86, 91,
102, 110, 157, 159–60,
160, 175, 184–5, 190,
195–6
Popov, Nikita 97
Providence, Rhode Island
154, 187
Ptolemaic planetary system
xxvii
Pushkin, Aleksandr 47

Queirós, Pedro Fernandes de
105

Richmond: Old Deer Park
observatory 102, 140*n*,
186
Rio de Janeiro 165–6, 196*n*
Rittenhouse, David 144–5,
151, 152, 153, 185
Rodrigues, island of 22, 24,
27, 28, 29, 66–7, 74–5,
87–8, 98
Royal Observatory *see*
Greenwich

Royal Society 13–14, 14;
and Delisle's proposal
14–18, 20; and war 19;
and Pingré 25, 32, 75,
87, 88, 91, 95; requests
ship from Admiralty
31, 33; and Mason and
Dixon 37, 39, 40–41,
58, 60, 63, 64, 91; and
first reports of 1761
transit 90, 91, 93; and
Lalande 96, 103*n*; and
Hornsby 105; plans
observation of 1769
transit 107, 108–9,
122, 123–5; obtains
funds from the Crown
126, 128, 140*n*; and
South Sea expedition
129, 131, 196*n*, 200;
request from Christian
VII to be made fellow
138; Benjamin Franklin
as fellow 145–6; and
Wales's Hudson Bay
expedition 148, 195;
sends team to North
Cape 156; publishes
North American results
195; hears British
results 200–1, 202;
Banks becomes presi-
dent 204; *see also*
Maskelyne, Nevil
Rumovsky, Stepan: views
1761 transit 83, 113;
battle with Popov 97;
in charge of planning
expeditions 114, 115,
116; has hope of

success for Russian
astronomers 136; his
assistant dies 182*n*,
191; views 1769 transit
119, 176, 191, 201
Russian Academy of Sciences
see Imperial Academy
of Sciences

St Helena 15, 16, 17, 31,
33, 34, 36, 36–7, 41,
61, 61–2, 80–81, 89,
90, 93
St John's, Newfoundland
81–2
Sajnovics, János 139–40,
161–4, 166–7, 181–2,
183–4; *Demonstratio
idioma Ungarorum et
Lapponum idem esse*
203 *and n*
San José del Cabo 176–7,
180–81, 189–90, 201
Scanderoon (Iskenderun),
Turkey 40
Seahorse, HMS 38–9, 40
Seven Years' War 3–4, 19,
38–9, 42*n*, 54, 67, 87,
101, 103, 108, 112, 157
Short, James 96, 98, 116,
118–19, 146
Silberschlag, Georg
Christoph 77*n*
Society of Jesus *see* Jesuits
Solander, Daniel 178, 180,
186*n*; results 202
solar parallax 70–71, 97, 98,
103, 201, 202
solar system: measurement of
distances xxvi, xxviii,

xxix–xxx, 5–6, 10, 11,
70–71, 97; *see also*
solar parallax
Solomon Islands 105
Stamp Act (1765) 102, 145
Swedish Royal Academy of
Sciences, Stockholm
xxv, 41, 52, 54–5,
78, 133–4, 136,
137, 205; *Kungl.
Vetenskapsakademiens
handlingar* 91
Swift, Jonathan: *Gulliver's
Travels* 33
Sylphide, Le (ship) 28, 71–2

Tahiti 6, 130, 167–70, *170*,
175, 177, *177*, 200,
201
Tasman, Abel Janszoon 105
telescopes xxv, xxix, 5, 25,
39, *53*, 55, *59*, 62, 74,
78, 82, *85*, 114,
118–19, 125, 139, 151,
180, *187*; achromatic
118, 160, 181
Tobolsk, Siberia 6, 8, 12, 43,
48, 49, 51, 60, 65, *65*,
68, 72–4; results 92, 96
Torneå, Lapland 41, 86,
104, 105, 134, 136
Tychonic planetary system
xxvii

Vardø, Norway 104, 123,
137–8, 140, 156, 161,
163–4, 169*n*, 181,
183–4, 195, 201;
observatory 163–4,
164; map *182*

Venus: transits xxiii–xxiv, xxviii, *xxix*, 5, 6, 17*n*, 69, 69–70, 70, 71, *84*, 103–5, 206; 'black drop effect' 92–3, 94, 118, *179*, 181, 192, *193*; luminous ring 192, *193*

Vienna: Royal Observatory 138

Voltaire 112, 117

Wales, William: and Hudson Bay expedition 127–8, 129–30, 141, 142, 147–8, 155–6, 175; successful results 195, 201

Wallis, Captain Samuel 130, 167 *and n*

Wargentin, Pehr Wilhelm 54; organises transit observations 41, 54–5, 57, 83; and Planman 55, 77, 78, 89, 94–5; views 1761 transit 78–9, 93; unsure about accuracy of observers' data 96–7; and organisation of 1769 expeditions 125, 133–6, 137, 138, 140, 175; and observation results 191, 193, 194

Wilkes, John 130

Winthrop, John xxvi, 56–7, 81–2, 146, 147, 153, 154, 176

Yakutsk, Russia 117, 119, 176

Andrea Wulf

The Brother Gardeners
Botany, Empire and the
Birth of an Obsession

Longlisted for the Samuel Johnson Prize for
Non-fiction 2008

'*Wondrous . . . I have learned so much from her book*'
JON SNOW, CHANNEL FOUR NEWS

One January morning in 1734, cloth merchant Peter
Collinson hurried down to the docks at London's Custom
House to collect cargo just arrived from John Bartram in the
American colonies. But it was not bales of cotton that awaited
him, but plants and seeds...

Over the next forty years, Bartram would send hundreds of
American species to England, where Collinson was one of a
handful of men who would foster a national obsession and
change the gardens of Britain forever: Philip Miller, author of
the bestselling *Gardeners Dictionary*; the Swede Carl Linnaeus,
whose standardised botanical nomenclature popularised
botany; the botanist-adventurer Joseph Banks and his
colleague Daniel Solander who both explored the strange flora
of Tahiti and Australia on Captain Cook's *Endeavour*.

This is the story of these men – friends, rivals, enemies, united
by a passion for plants. Set against the backdrop of the emerging
empire and the uncharted world beyond, *The Brother Gardeners*
tells the story of how Britain became a nation of gardeners.

'*Absorbing and delightful*'
JENNY UGLOW, SUNDAY TELEGRAPH

ANDREA WULF

The Founding Gardeners
How the Revolutionary Generation
Created an American Eden

The New York Times Bestseller

A follow-up to Andrea Wulf's award-winning and critically
acclaimed history of British gardening, this is the story of how
George Washington, Thomas Jefferson, John Adams and James
Madison's passion for nature, plants, agriculture and gardens
shaped the birth of America.

Through a series of vignettes spanning the Declaration of
Independence to the death of Adams and Jefferson exactly fifty
years to the day afterwards, these stories weave the political,
the personal and the botanical and are in turns funny, fascinating and moving. The Founding Gardeners shows that it is impossible
to understand these visionary men and the American nation
without considering their love of gardening.

'Engrossing . . . excellent . . . fascinating . . . a timely and passionate book'
GUARDIAN

'Superb . . . this book will fascinate anyone interested in gardening,
agriculture or American history'
MAIL ON SUNDAY

'Wonderfully engaging . . . Her knack for description is marvellous'
TIMES LITERARY SUPPLEMENT